Make:

Getting Started
with Adafruit Circuit
Playground Express

**THE MULTIPURPOSE LEARNING
AND DEVELOPMENT BOARD
WITH BUILT-IN LEDS, SENSORS,
AND ACCELEROMETER**

Anne Barela

Foreword by Limor "Ladyada" Fried

Make Community, LLC
Santa Rosa, CA

Published by
Make Community, LLC
150 Todd Road, Suite 200
Santa Rosa, CA 95407

Make books may be purchased for educational, business, or
sales promotional use. Online editions are also available for most titles
(safaribooksonline.com). For more information, contact our corporate/
institutional sales department: 800-998-9938 or *corporate@oreilly.com*.

Publisher: Roger Stewart
Editor: Patrick Di Justo
Copy Editor: Elizabeth Welch
Proofreader: Scout Festa
Interior and Cover Designer and Compositor: Maureen Forys,
Happenstance Type-O-Rama
Indexer: Valerie Perry, Happenstance Type-O-Rama

September 2018: First Edition

Revision History for the First Edition

2018-09-15 First Release

2020-04-03 Second Release

See *oreilly.com/catalog/errata.csp?isbn=978-1-68045-488-8* for release details.

978-1-68045-488-8

O'Reilly Online Learning

For more than 40 years www.oreilly.com has provided technology and business training, knowledge, and insight to help companies succeed.

Our unique network of experts and innovators share their knowledge and expertise through books, articles, conferences, and our online learning platform. O'Reilly's online learning platform gives you on-demand access to live training courses, in-depth learning paths, interactive coding environments, and a vast collection text and video from O'Reilly and 200+ other publishers. For more information, please visit www.oreilly.com.

Please address comments and questions to the publisher:

Make Community, LLC
150 Todd Road, Suite 200
Santa Rosa, CA 95407

You can send comments and questions to us by email at books@make.co

Make: Community is a growing, global association of makers who are shaping the future of education and democratizing innovation. Through *Make:* magazine, 200+ annual Maker Faires, *Make:* books, and more, we share the know-how of makers and promote the practice of making in schools, libraries, and homes.

To learn more about Make: visit us at *make.co.*

Foreword

The story of Circuit Playground begins maybe eight years ago. Adafruit was still an apartment company then. My partner and I were chatting with a middle school superintendent who told us that the school was being pitched STEM (science, technology, engineering, and mathematics) education products for its students (the products were similar to tablets with sensors that could plug in). But at $500 each, the school could afford only one per classroom. So twenty-plus kids would have to share.

At the time, Arduino was becoming popular—it's a lot less expensive! But younger students struggled with learning C++ (especially if they were coming from block-based Scratch programming), and the setup could get complicated since Arduino requires a special development environment.

For a long time, I didn't have a solution to these problems. The slick technology was just too expensive, and the low-cost educational kits were too hard to use. But eventually, enough stuff was invented (low-cost ARM Cortex microcontrollers! NeoPixels! Embedded Python!) that we were able to make the ultimate circuit board for teaching coding and electronics.

That's where Circuit Playground comes in. Easy to use, fun to program, and affordable for any student, it works with the Mac, Windows, Linux, Chrome OS, and even Android! You can use it at home, at school, at work, or on a library computer—no software needs to be installed.

We poured all the know-how and experience we've had over 10 years of selling educational electronics to create something for everyone. Whether you want to build cosplay props, scientific experiments, robotics, or spy gadgets—in drag-n-drop Microsoft MakeCode, interpreted CircuitPython, or Arduino—Circuit Playground Express will be your companion as you learn and create.

—*Limor "Ladyada" Fried, founder and engineer, Adafruit*

Preface

Adafruit Circuit Playground Express provides a low-cost way to explore programming, sensing, and interaction. The Express is a microcontroller-based electronics and software development board. It is programmable in Microsoft MakeCode, JavaScript, and Python and with the Arduino development environment. Its built-in motion, temperature, and light sensors let Circuit Playground Express sense the world around it. Its 10 NeoPixel lights and speaker allow Circuit Playground Express to communicate with the outside world.

Circuit Playground Express is different from many beginning electronics available today. Out of the package, Circuit Playground Express can be connected to a computer that runs any operating system. Load Microsoft MakeCode in an Internet-connected web browser, and in less than 15 minutes you'll have an interactive project all your own.

Think—would you fancy clothes or shoes with LEDs that dance to movement and music? Would you like a musical synthesizer that plays your choice of sounds, even using fruit as your input? Perhaps a light-up pin that makes Star Trek–like sounds when tapped? All these and many, many more can be built using Circuit Playground Express right out of the package!

This book provides the information to get you started using Circuit Playground Express quickly. The information and ideas in the book may be the foundations for your own projects and explorations.

WHO THIS BOOK IS FOR

This book is for the enthusiast, the student, the curious person who wishes to expand their knowledge of making through interactivity, sensing, lights, or sound.

Skills that are useful in working through this book:

* A knowledge of the fundamentals of what software and hardware are.

* Experience with desktop or laptop computers running an operating system such as Microsoft Windows, Apple macOS, Chrome OS/Chromebook, or Linux. Skills include navigating a filesystem and selecting specific files to use.

* Use of a graphical Internet web browser. Many are available, including Chrome, Firefox, Safari, Internet Explorer, and Microsoft Edge.

* Use of a text-based editor on one of the listed operating systems and the ability to open a text file, change the file, and save the file both to the computer disk and to a flash drive connected to the computer.

Working with Circuit Playground Express is suitable for beginners who do not know electronics or programming. After you finish reading, you can use this book as a reference for the techniques presented.

PREPARATION

There is no required reading to work with this book, but here are some suggested resources that you may draw on to better understand particular subjects as the book progresses.

MakeCode

The Microsoft MakeCode.org website (*https://makecode.com/#learn*) is a good reference. Adafruit has a free tutorial on learning Makecode (*https://learn.adafruit.com/makecode*). Adafruit continually publishes new projects and tutorials on Circuit Playground Express at learn.adafruit.com (*https://learn.adafruit.com/*). Finally, Adafruit has support forums for assistance at forums.adafruit.com (*https://forums.adafruit.com/*).

Python Basics

The website python.org (*www.python.org*) provides free materials (*www.python.org/about/gettingstarted/*) to help you learn the Python programming language.

Arduino

The book *Getting Started with Arduino, Second Edition*, by Massimo Banzi (co-creator of Arduino), is a good resource to start with. I also recommend the Adafruit Learn Arduino series (*https://learn .adafruit.com/lesson-0-getting-started*), available for free online. Both offer an introduction to the Arduino open source electronics prototyping platform, including programming.

WHAT YOU WILL WANT TO HAVE ON HAND

To program Circuit Playground Express, you will need a Windows, Mac, or Chromebook computer with a USB port. You need Internet access to run the Microsoft MakeCode editor and to download example code, rather than typing it in yourself.

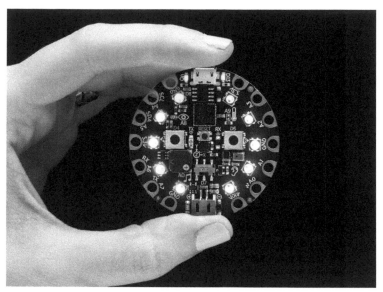

FIGURE 1-1. Circuit Playground Express from Adafruit Industries

You may have heard of some of the devices behind the proliferation in personal programmable electronics. The Arduino and Raspberry Pi are probably the best known; the micro:bit is also being introduced into classrooms. These and other devices often require a good number of external components to do much beyond basic demonstrations. There are few learning boards at an affordable price that do not require extra parts.

Enter Limor "Ladyada" Fried (Figure 1-2). She studied engineering at the Massachusetts Institute of Technology (MIT). Starting in her dorm room, she published electronic projects on the Internet for free. People started asking if she sold the parts used in her designs, and she did so, founding the company Adafruit Industries, LLC. She moved the company from MIT to Manhattan in 2005, and Adafruit has grown due to Ladyada's popular lineup of information and electronic products.

FIGURE 1-2. Limor "Ladyada" Fried, engineer and founder of Adafruit

Limor likes to call Adafruit "a learning company that sells electronics" rather than a company that just provides parts and generic support. Adafruit has well over fifteen hundred tutorials on the Adafruit Learning System at *https://learn.adafruit.com/* to guide people on a wide range of electronic projects using its product line. Adafruit provides guides under the Circuit Playground name. This includes electronic characters (see Figure 1-3), videos, and easy-to-use electronics.

FIGURE 1-3. The Adafruit Circuit Playground characters

The electronics beginner products are near and dear to the Adafruit mission. They answer the question, "Can we produce an affordable device, allowing people to learn and interact quickly and easily?"

The result is the Circuit Playground line of products, which were designed from the start to be user friendly and packed with features, at a price any hobbyist or classroom can afford. The first board was Circuit Playground Classic (Figures 1-4 and 1-5, left). Later, Adafruit refined the design, creating Circuit Playground Express (Figures 1-4 and 1-5, right). The Express is the focus of this book.

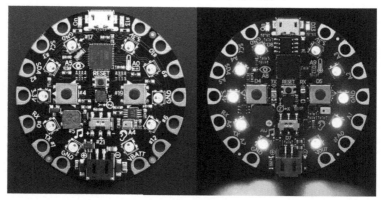

FIGURE 1-4. Circuit Playground Classic (left) and Circuit Playground Express (right)

FIGURE 1-5. On the back, identifying Circuit Playground Classic (left) and Circuit Playground Express (right)

Circuit Playground Express is tiny—only 2″ (5cm) in diameter. This small space is filled with miniature electronics goodness. Fourteen inputs and 20 outputs provide a number of ways to sense and interact. At its heart is the powerful yet easily programmable Microchip ATSAMD21 microcontroller. A small flash chip enables Circuit Playground Express to act like a USB flash storage drive (sometimes called a thumb drive). It can be powered from a USB cable connected to a computer or via an attached battery pack. The board's capabilities easily allow for many different types of projects.

The Circuit Playground Express ease of use extends to programming as well. Programming sounds complicated, and it was in early electronics design. But modern products such as Circuit Playground Express are designed with ease of coding baked in. Circuit Playground Express can be programmed in Microsoft MakeCode, an environment similar to the Scratch programming language. The board can also be programmed with CircuitPython, an onboard subset of the popular Python language. Finally, Express may be programmed via the Arduino Integrated Development Environment (IDE), using the C and C++ programming languages along with hundreds of existing public domain code libraries. All three methods of programming are available at no cost.

Freedom is part of the Circuit Playground Express design. Adafruit freely provides all the information available to understand and build the board through open source. Open source is a movement where hardware and software are freely available for study or reuse. The spirit of open source hardware and software has been adopted by a large number of people and companies worldwide, accelerating the ability to design new things by utilizing concepts from other open source projects. And this concept enables us to study and understand how Circuit Playground Express functions, allowing for a better understanding of what is going on "behind the scenes."

Throughout this book, you will learn the various ways to interact with a Circuit Playground Express. The examples will demonstrate the basics. Readers are encouraged to go beyond the examples to build their own projects and to realize their own creative ideas.

In the next chapter, we'll look in-depth at the capabilities of Circuit Playground Express, plug it in, and "kick the tires."

2

. . . .

A Tour of Circuit Playground Express

et's look at Circuit Playground Express in detail. On the front you'll see the features outlined in Figure 2-1.

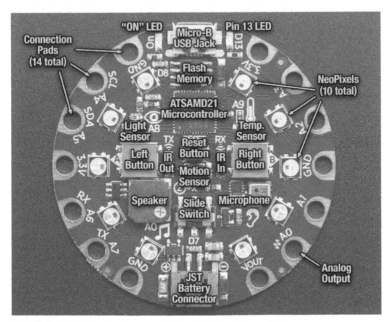

FIGURE 2-1. The features of Circuit Playground Express

The microcontroller is the heart of the board. It processes input, makes detailed calculations, and presents outputs.

As of this writing, the chip used on Circuit Playground Express is the ATSAMD21G18A-AU manufactured by Microchip Technology (formerly Atmel). This processor has more features than the typical Arduino and runs at a speed eight times faster than the Arduino Uno. The extra processing capability of the microcontroller allows for advanced features such as multiple inputs and outputs, true analog output, Universal Serial Bus (USB) communications with a host computer, and hardware-assisted capacitive touch control.

For the technically minded, the microcontroller is based on the ARM Cortex-M0+ running at 48MHz. It has 48 input/output pins, 256KB of flash memory, and 32KB of static random access memory (SRAM). Some of the flash memory is used by a bootloader program to allow the user to easily interact with Circuit Playground Express, and the rest is available for user code.

As faster processors become available at affordable prices, it is conceivable that Adafruit may substitute a new chip for the ATSAMD21 or otherwise enhance the design. But Adafruit has assured customers that the functionality of any revised Circuit Playground Express will be the same to the user. Newer boards may have more flash memory, operate faster, or have a feature or two in addition to the capabilities we'll discuss, but all the programs and concepts in this book will remain on point for years to come.

The flash memory chip works like a *flash drive* (or a thumb drive or SD card) to store files when using CircuitPython. Space is small by modern standards, only 2MB, but this is usually plenty for a Python program. This chip may also be accessed via the Arduino programming environment, but it takes more code than to access it via CircuitPython. Microsoft MakeCode also stores its programs in the flash memory.

Best Advice in This Book: Get a Good USB Cable

USB A male–to–USB micro B cables may be available in most households due to the proliferation of electronics in our lives. But not all USB cables are made the same. Some have power wires to recharge devices only, often found with external "cell phone recharging" batteries. Such power wiring–only cables will not be able to make data connections, meaning you cannot use that cable to program your device. Other cables may be of poor quality or stressed in some way to make poor connections. Flexing between the connector and cable makes wiring fray and become intermittent.

When you get your Circuit Playground Express, I strongly suggest that you also get a new, compatible USB cable to use with your projects. Adafruit sells good cables as product ID 592. Good cables are also available from mobile phone outlets, electronics shops, and online. Again, spend a bit extra to ensure the USB cable purchased is a quality cable; the lowest price on eBay may not buy the reliable cable you hope for.

The USB port is typically used to connect your Circuit Playground Express to a computer to program. The connector is USB micro B, the same connector used in many consumer electronic devices, including non-Apple phones and tablets.

The USB connector can also be used for communications in your projects, similar to other USB devices you use with a computer. The port is also the default text output device in Circuit-Python and Arduino programs. You can use a terminal emulator application on your computer to see messages from USB.

Batteries—On the Go!

Using a battery pack allows Circuit Playground Express projects to be taken nearly anywhere! Here are the suggested methods for safely powering the board.

From the USB port, you can plug in a USB cable from your desktop or laptop computer. Mac, PC, or Linux computers work great for USB power.

A 5-volt "cell phone external battery pack" such as the models available at Adafruit (*www.adafruit.com/ ?q=battery%20pack*) works great connected to the Circuit Playground Express USB port.

If you design a power-hungry project using external components, such as a large number of NeoPixels or other lights, you should use either a wall/mains-powered 5-volt power supply or a larger battery pack.

When using AA or AAA batteries, it is best to consider a 4.5-volt pack using three cells. You can connect these to the proper JST connector for the board or buy preassembled battery packs. Adafruit sells the following models:

- ❋ 3 x AAA Battery Holder with On/Off Switch and 2-Pin JST, Product ID: 727
- ❋ 3 x AAA Battery Holder with On/Off Switch, JST, and Belt Clip, Product ID: 3286
- ❋ 3 x AA Battery Holder with On/Off Switch, JST, and Belt Clip, Product ID: 3287

A smaller, coin cell holder is available as the 2 × 2032 Coin Cell Battery Holder – 6V output with On/Off switch, Adafruit Product ID: 783. Though smaller, this pack will not power a hungry circuit for a long time, and the coin cells are more expensive than AA and AAA batteries.

The circuitry on the board will use either the battery input power or USB power, safely switching from one to the other. If both sources of power are connected, Circuit Playground Express will use whichever input has the higher voltage. The device works great with a lithium-polymer battery or the Adafruit 3 × AAA battery packs with a JST connector on the end. The connector allows for swapping batteries if they run out of juice.

There is no built-in battery charging circuitry in Circuit Playground Express, which means that you can use alkaline or lithium batteries safely. Unfortunately, that also means rechargeable batteries must be recharged by removing them from the Circuit Playground Express project and charging them in the manufacturer's approved charger.

What Is a LiPo battery?

Lithium-polymer (LiPo) batteries are often used in wearable or portable projects. Their thin size and rechargeability make them very popular with experimenters. They come in multiple sizes (see the following graphic); the larger batteries last longer.

Several Different LiPo Battery Sizes

With the correct JST connector as used on Circuit Playground Express, you can get LiPo batteries in sizes from 350 milliamp-hours (mAh) to a whopping 6600mAh, with the larger sizes thicker and heavier.

A program not using NeoPixels may use less than 20 milliamps (mA). Lighting all 10 NeoPixels, as in the demonstration program, may draw up to 30mA. Using all NeoPixels at maximum brightness will draw more current, but doing this is rarely required.

To get an estimate for runtime from a battery, divide the capacity of the battery by the battery rating. For a small 150mAH battery, 150 mAh / 30 mA = 5 hours.

The time might be for a perfect, fully charged battery. If you plan to depend on a period of operation for a project, consider testing the battery capacity beforehand. Then consider carrying a spare battery for actual use.

A selection of batteries and chargers are available on the Adafruit website (*www.adafruit.com/category/916*).

That's our tour of Circuit Playground Express. One more thing: On the back of the board, there is a place to write your name! Identifying the board as yours may be handy in a classroom environment or if you're working with friends. That space can accommodate some tape to write your name if the board is being borrowed.

Use Only the Manufacturer's Battery Chargers

It is vital that you use only the chargers provided by the manufacturer. Using other charging methods for batteries, especially lithium-based batteries, may cause fires (this is why they are often banned from airplanes).

With proper charging and use, LiPo batteries are safe and convenient.

But if you are unsure, stick to AA or AAA battery packs for light to moderate use and D cell packs for heavy or prolonged use lasting many hours.

OPERATING SYSTEM SOFTWARE SETUP

Fortunately, no software driver installation is needed for the following operating systems. You can safely proceed to the next chapter.

* Windows 10 (if you have Windows 7 or Windows 8, keep reading)

* macOS

* Linux

* Chromebook

NOTE You need to install drivers only for Windows 7 and Windows 8. Windows 10 and non-Windows operating systems have drivers built in.

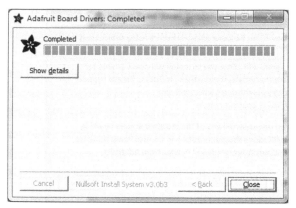

FIGURE 2-5. The progress bar and completing the driver installation

That's it! If you're using Windows 7 or 8, you should be set to plug in your Circuit Playground Express and start working.

CHAPTER QUESTIONS

1. How many physical switches are on Circuit Playground Express?

2. Can a Circuit Playground Express light up in orange or blue? Which parts, if any, can do this?

3. How might you power Circuit Playground from a vehicle?

3

Getting Started with Microsoft MakeCode

Enough background information! It is time to plug Circuit Playground Express into a computer and get a feel for what it can do.

If you have not had a chance to get a Circuit Playground Express yet, you can still go to the Microsoft MakeCode section and start to program, because the MakeCode environment has a Circuit Playground Express simulator to show you what the code will do. You can then save that code and run it on Circuit Playground Express when you are able to get one. Figure 3-1 shows the MakeCode logo.

FIGURE 3-1. The MakeCode logo (Credit: Microsoft)

Next, plug the USB "full-size" A connector into a USB socket on your computer.

Wow!

Did that scare you? It did me when I first tried it! The default program on Circuit Playground Express runs as soon as the device has power. The program from the factory blinks the 10 NeoPixel lights and plays tones.

If your Circuit Playground Express does not appear to do anything, the board may have been used before, which would likely remove the factory demonstration program. No worries! You should see the green "On" LED lit at a minimum, letting you know it has power.

> NOTE You can silence the demonstration program by moving the slide switch tab toward the speaker. The lights will continue to flash but the sound will stop.

With an indication Circuit Playground Express is connected to the computer, it is time to write some code!

MAKECODE: YOUR FIRST PROGRAM

If you have never developed computer code in your life or you have some coding skills in another programming language, you should try MakeCode on your Circuit Playground Express.

No software installation is required: MakeCode (*https://makecode.com/*) runs in any modern web browser on most operating systems. Once the web app has been loaded from the Internet, all its features will continue to work, even if you disconnect the computer from the Internet.

MakeCode provides the following:

* A block editor (similar to the Scratch programming environment) and a JavaScript editor to create programs, with the ability to convert back and forth between visual and text-based program representations

* A web-based simulation of the physical device (Circuit Playground Express, micro:bit, and several other development products are supported) so that students can edit and test their programs, even if they don't have a device (or if they left it at home or school)

* A self-guided "Getting Started" experience to introduce the basic features of the programming environment, as well as a set of projects for making and coding

* A compiler that instantaneously creates an executable file to download/copy to Circuit Playground Express

* A sharing feature so that students can share their programs with students, teachers, parents, and friends

MakeCode also adapts to the screen size of your computer; it works well on desktops, laptops, tablets, and even smartphones. For this introduction, let's assume you are using a desktop, laptop, or Chromebook.

Start your web browser software. The web browser software should be fairly modern and capable, such as Chrome, Edge, Safari, or Firefox. In your browser, type `https://makecode.adafruit.com/` into the address bar and press Enter. You should see a screen fairly similar to the one in Figure 3-4. The background image and the screen size may be a bit different, but that's fine.

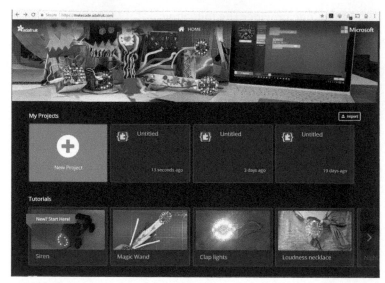

FIGURE 3-4. The Google Chrome browser opening MakeCode

Click the blue New Project square under My Projects. You should get the screen shown in Figure 3-5.

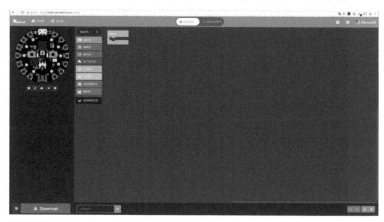

FIGURE 3-5. The MakeCode screen for Circuit Playground Express

The Circuit Playground Express on the screen in the upper left is a functioning simulator. Anything done to program Circuit Playground Express can be simulated in MakeCode.

Look at the screen in Figure 3-6. There's a picture of a Circuit Playground Express in the upper-left corner. You can see a rainbow stack of command blocks running vertically down the center, and there's a green block shaped sort of like the letter C, with the word *forever* on it.

FIGURE 3-6. A close-up of the MakeCode screen

NOTE Your computer no longer needs an Internet connection once you start MakeCode. MakeCode will continue to run in your web browser. The sharing portion of the environment will not work, but the coding environment will. When you later reconnect to the Internet, the MakeCode environment will fully work with the sharing feature. This capability is great if you take your computer on the go. An Internet or WiFi connection is not required to continue to use MakeCode.

The colored blocks group is to the right of the Circuit Playground Express replica and below the search box. These buttons provide different functional code blocks to use in coding. If you click one, MakeCode will show you the group of commands. For example, clicking the LIGHT button "pops out" to the right various code blocks related to working with the 10 NeoPixels around the board (see Figure 3-7).

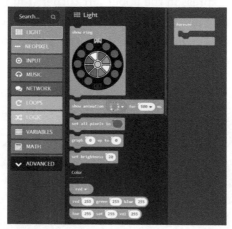

FIGURE 3-7. Clicking the LIGHT button pops out light-specific programming blocks.

You can click each of the buttons representing functional groups to get a feel for all the code blocks available for Circuit Playground Express in MakeCode. The available blocks may change from board to board—Circuit Playground Express has a great number of blocks to take advantage of all the functionality built into the device.

At this point, let's jump right in to using MakeCode. With your mouse, move to the blue block under the LIGHT heading marked set all pixels to that has a red circle next to it. Hold down your left mouse button and move the mouse to the empty part of the forever green letter C–like block. Release the mouse button. Your screen should look like the one in Figure 3-8. If the blocks did not appear to "click together," move the blue block so that it interlocks with the green block.

FIGURE 3-8. Moving the `set all pixels to` block into the green `forever` block

You have just completed a functional program! What is this code doing?

The `forever` block is a code loop that says "examine all the blocks between the top and the bottom of the green area, and do those command(s) over and over, one after another, forever" (hence the block name *forever*).

You will note the Circuit Playground Express simulator now has all the NeoPixels set to red. The simulation is already running your code! Congratulations, you have written your first MakeCode program and it is running in the simulator. Just like that!

The color of the `set all pixels to` block is red by default. Default means that it's the command or color or value that will be automatically chosen when the block is used. You could change the color by clicking the red circle and selecting a different color.

Let's change things. On the blue `set all pixels to` block, click the red circle and click blue. Immediately, all the pixels in the Circuit Playground Express simulator turn blue (see Figure 3-9)! Feel free to select different colors. NeoPixels can be set to tens of thousands of colors. Change the color to blue when done so you know what color you set here.

FIGURE 3-9. The program that sets all NeoPixels to blue

Next we'll program your physical Circuit Playground Express. It takes just a couple of steps to put the code on the board and have it run. Note that this is not required if you have not yet acquired a board yet—you can skip the next section and come back later.

UPLOADING MAKECODE TO CIRCUIT PLAYGROUND EXPRESS

Though simulating code in the MakeCode environment gives you a feel for what the commands are doing, there is a satisfying knowledge when that code actually does something in the physical world.

The steps that follow outline how to send the code to a Circuit Playground Express board that is connected to the computer via a USB connection. The explanation appears a bit long, but the steps become second nature after you practice a time or two.

There are only two steps to download your completed Make-Code program:

Step 1: Connect your board via USB and enter bootloader mode.

Step 2: Compile and download the UF2 file into your board drive.

We will go through these two steps in detail.

The UF2 File Format

Every file on a computer is structured or formatted to be recognized as containing data for a specific purpose. For example, files with the extension .tif are JPEG compressed images/pictures, and files with the extension .doc or .docx are Microsoft Word files.

Microsoft has created a file format for use with microcontrollers like Circuit Playground Express. A UF2 file provides information in a basic format that can easily be recognized by a microcontroller. If you are curious, the specifications on what a UF2 file is and code that makes use of UF2 files are here: *https://github.com/microsoft/uf2*.

UF2 can be used for two different types of files. The first is for bootloaders, the basic code placed on a microcontroller to instruct it how to do basic tasks and read/write to devices connected to it. The second type of file is for code files placed in a filesystem.

For MakeCode, the code placed on Circuit Playground Express will be in the UF2 file format.

Step 1: Bootloader Mode

Connect your board to your computer via a USB cable. Press the Reset button once to put the board in bootloader mode (Figure 3-10).

NOTE If this is your first time running MakeCode or if you have previously installed Arduino or Circuit-Python, you may need to press the Reset button twice to get your board into bootloader mode.

FIGURE 3-10. Pressing the Circuit Playground Express Reset button to get into bootloader mode

When Circuit Playground Express is in bootloader mode, all 10 NeoPixel LEDs will turn red briefly, and then turn green. Verify that your status LED (marked "D13," next to the USB jack) is also pulsing red.

> **NOTE** If the NeoPixel LEDs all stay red after you press the Reset button, either the computer is still installing drivers (please wait a minute; Windows takes some time to install updates) or you have a bad USB connection. If you keep getting red NeoPixels, try a new USB cable.

You may want to ensure your USB cable is not "charge only." The cable needs to be a fully functional USB cable that can transfer data. You may also need to use a different USB port on your computer, or a port that is not connected to a USB hub of some sort. USB errors are not intuitive, but swapping things around will usually result in a good connection, as long as you have a good cable from the host computer to Circuit Playground Express.

Once your LEDs are all green, you should see a drive named CPLAYBOOT appear in your drive list in your operating system file explorer (see Figure 3-11).

FIGURE 3-11. The CPLAYBOOT drive in Windows Explorer (left) and in the Mac Finder (bottom right)

This is the onboard "thumb drive/flash drive." The flash drive is the space to which you copy your program files. If you are not familiar with the files portion of your computer, you may want to have someone come over and show you where the CPLAYBOOT drive is located.

We are now ready to compile our code and download it to the board!

Step 2: Compile and Download

Let's first verify that our code compiles properly in MakeCode.

MakeCode has a built-in simulator that reloads and reruns code when restarted. This is an easy way to ensure that our code compiles and to simulate it before moving it onto the board.

There are five button icons below the image of Circuit Playground Express on the MakeCode screen (see Figure 3-12). From left to right they are as follows:

* Stop: Stops the code from running

* Refresh: Restarts the code running

* Step: Goes through a program step by step

* Audio: Turns sound on and off

* Full Screen: Toggles the simulation in a full-screen window

The refresh button reloads the simulator with your latest version of block code.

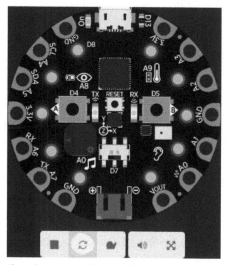

FIGURE 3-12. The simulated Circuit Playground Express with the five control buttons. The refresh button is highlighted in yellow.

> **NOTE** What if a message pops up when you click the refresh button? If you receive a "We could not run this project" error, please check over your code for errors. For the simple program that lights the NeoPixels, there should not be a message. If you code a more complicated program, this message lets you know to recheck the program steps in the code blocks.

If your board is working in the simulator, it's time to download it to your actual Circuit Playground Express. Click the Download button (Figure 3-13). It will generate a UF2 file and then download it to your computer.

FIGURE 3-13. The Download button

A nice graphic explains moving a file from your computer file explorer to the Circuit Playground Express CPLAYBOOT flash drive (Figure 3-14).

FIGURE 3-14. When the code download is complete, these steps are displayed.

1. Ensure your board is in bootloader mode by pressing the Reset button once (or twice if necessary). When in bootloader mode, all the NeoPixel LEDs are lit green.

2. Find the UF2 file generated by MakeCode in your file explorer. Copy it to the CPLAYBOOT drive.

3. The D13 LED on the board will blink while the file is transferring. Once it's done transferring your file, the board will automatically reset and start running your code, just like in the simulator.

Windows

1. Open Windows File Explorer (Windows key + E key) and locate the `circuitplayground-Untitled.uf2` file you generated (or whatever name you saved it as). It's probably in your `Downloads` folder unless you saved it somewhere like the desktop, a personal flash drive, or a program save folder.

2. Either copy and paste the file to your CPLAYBOOT volume or drag and drop it as shown in Figure 3-15.

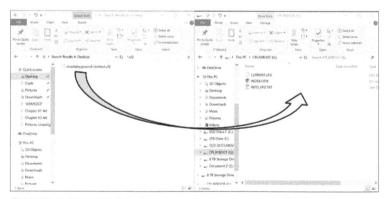

FIGURE 3-15. Two side-by-side File Explorer windows, one containing the program UF2 file, the other the contents of the CPLAYBOOT drive on Circuit Playground Express

macOS

* Open Finder and locate the file named `circuitplayground-untitled.uf2`. You can copy and paste this file to the CPLAYBOOT volume or drag and drop it from the same Finder window.

> NOTE If you want to avoid the copying process, you can download your programs directly to the board. To do this, change the download location in your web browser program to the main directory of your CPLAYBOOT drive.

Chromebook

1. Your Circuit Playground Express will be recognized by your Chromebook. All the NeoPixels will be lit green when it is ready to download a file. If the NeoPixels are not lit, press the Reset button once to put the board into flash drive mode and turn the NeoPixels green. A small pop-up will notify you that a new drive was detected called CPLAYBOOT.

2. Click the Download button. In the window that appears, click CPLAYBOOT on the left. You may change the filename of your program, listed at the bottom of the pop-up window, from `circuitplayground-Untitled.uf2` to anything you like, but be sure it ends in `.uf2`.

3. Click Save. The red D13 LED on the board should blink red as the program transfers (this might happen too quickly for you to see). Circuit Playground Express will reboot after the download, on its own, and your program should now be running.

Done!

The program is running; the NeoPixels are lit. Your program should run as soon as Circuit Playground Express detects that you have saved a valid UF2 file to the onboard flash. For the previous program that turned all NeoPixels to blue, the physical Circuit Playground Express on the desk should have all the NeoPixels displaying blue, unless you selected another color when you were done. That is *your* code running on Circuit Playground Express.

Congratulations, you just programmed a computer!

In the off chance the board did not reset itself, you can press the Reset button twice (the first press places it into bootloader mode, the second tells it to run current program mode). You will want the NeoPixel LEDs all green and a flash drive called CPLAY-BOOT to show up on your computer. If you believe you are having problems, check Appendix A for issues and advice.

What Happens if the Power Is Removed?

All the programming methods you can use on Circuit Playground Express store code in flash memory. Just like a thumb/flash drive, the program will stay on Circuit Playground Express even though the power is removed. No battery is needed to keep the program on the board and ready for its next use.

This is in contrast to large computers where, if you turn off the computer, the current program is no longer in memory and the operating system is probably upset that it did not get to shut down gracefully!

You should save all your code on a backup drive such as a second thumb/flash drive or a hard drive. If by some means you were not able to save your last program on your computer, you can plug in your Circuit Playground Express via a USB cable and copy the last program back onto the computer. If you are in a classroom setting, your instructor may provide more information on saving your programming.

MODIFYING A PROGRAM

The first program was simple in order to get to a level of comfort with MakeCode. Next we'll modify the program to do something more dynamic.

Go back to your MakeCode program browser window. It should look as we left it, with the first program written in the coding space (Figure 3-16).

FIGURE 3-16. Our first program revisited

Click the blue set all pixels to block with your left mouse button and drag the block away from the forever block (Figure 3-17).

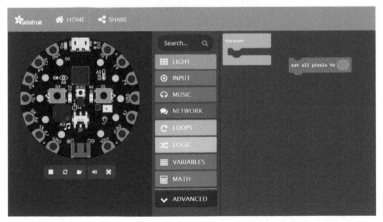

FIGURE 3-17. The set all pixels to block is moved outside the forever loop block.

Now let's make some better-looking lighting effects using another LIGHT group block. Click the LIGHT code block group button and look for the show animation blue block (Figure 3-18). Drag that block (Figure 3-19) into the forever block where the set all pixels to block had been (Figure 3-20).

FIGURE 3-18. Clicking the LIGHT button brings up a list of blocks related to LEDs. The show animation block is the second one down the list.

FIGURE 3-19. The show animation block

You can get rid of the set all pixels to block by right-clicking it (to highlight it) and pressing the Delete key (Del) on the keyboard. The block should disappear from the coding space.

What will the code do with the show animation block inserted? A quick look at the simulator will show you that the lights are now multicolored. The forever loop keeps the animation running indefinitely. The multicolor animation is popular, and it can be used for many projects.

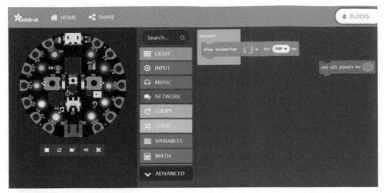

FIGURE 3-20. The `show animation` block placed into the `forever` loop block

You can download this program to your Circuit Playground Express if you wish. But changing a couple of things may be helpful.

There are two things you can change on the `show animation` block. If you click the triangle next to the colored wheel, a list of preprogrammed animations will pop up. This allows you to set different animations. Feel free to select each animation, and then see what the simulator will do. Set it back on "rainbow" when you are done.

The second changeable item on the `show animation` block is the number of milliseconds (ms) to perform the animation (1ms = 1/1000 of a second). This can range from 100ms (a tenth of a second) to 2000ms (2 seconds). If you change this value at this point, does anything happen?

No; as the `forever` loop immediately starts the animation after the time you set, it does not appear to affect the single animation. To show how we can add some time to the `forever` loop, we will click the green LOOPS block group, select the green `pause` block, and place it under the blue `show animation` block so that you have a program that looks like the one in Figure 3-21.

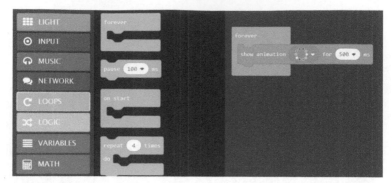

FIGURE 3-21. Clicking the green LOOPS block shows several code blocks. Select pause, left-click, and drag the block under the show animation block.

When you drag the pause block under the show animation block, the forever block will "open wider" to allow the pause block to fit inside (Figure 3-22).

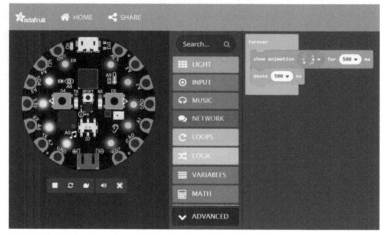

FIGURE 3-22. The pause block placed inside the forever loop under the show animation block

Change the time value on both the show animation and pause blocks to 500ms—500 milliseconds is 0.5 (one half) second. Those times are slow enough to see with the eye.

The program will now perform the rainbow animation, wait 500ms (half a second), and then show the rainbow animation again, repeating continuously. The simulator does not appear to show anything different. At this point, we'll add a second show animation block from the LIGHT block and another pause block from the LOOPS block. Make a stack within the forever loop that contains show animation, pause, show animation, then pause, as shown in Figure 3-23.

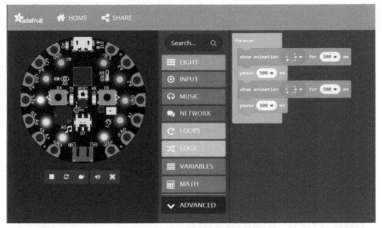

FIGURE 3-23. The program with two show animation and two pause blocks

The rainbow effect spins, then pauses, spins, then pauses on the simulator. On the Circuit Playground Express board, you might not even see the pauses. It is time to change some of the values. Change the time values for the animations to 2 seconds on each show animation block by clicking the black triangle next to the number 500 and selecting the value 2000ms. It is still simulating a rainbow. Time to change the second animation to something different, say the purple comet. Your screen should look like the one in Figure 3-24.

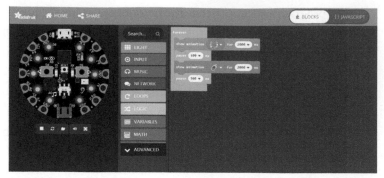

FIGURE 3-24. Two different animations, each running for 2 seconds

What is the new code doing? It loops forever, first displaying a rainbow animation for 2 seconds, waiting half a second, then displaying a changing purple animation for 2 seconds, then waiting half a second before starting all over again. This is a nicer effect than all red in the first program.

Using the instructions for downloading a program to an actual Circuit Playground Express, download this code to the board. Does it behave the same way as in the simulator?

Your Turn: MakeCode

Adjust the types of animation. How many animations does Make-Code provide? The time each animation runs can be changed. What happens when you make it rather short? The pause time can also be changed; how does that change the action? This experimentation will give you an idea what each part of the program is doing.

SAVING A PROGRAM

At any time, you can save your MakeCode to a file on your computer. When you decide to save a program, use the blue box next to the pink Download box. Type a descriptive name for your program (I called it **two animations**) and click the disk icon (see Figure 3-25).

FIGURE 3-25. The file save box is blue, next to Download. Type a filename and click the disk icon to save.

The operating system should show a Save As box that looks like the File Explorer (Windows) or Finder (Mac). On Chromebook, you will see a screen that lets you save the file to a number of places like Google Drive or to a disk drive such as a personal flash drive if you have one plugged in. You can save the file anywhere you like. You may want to create a directory called **Programs** (or another name you will remember). Your instructor may tell you where to save your programs if you're in a class.

If you are using a shared computer, consider saving the program on a thumb/flash drive. The drive doesn't not need to have a large capacity, since all the programs in this book are rather small files.

Where Did the Demonstration Program Go?

Once you load a UF2 code program that replaces the file current.uf2 on Circuit Playground Express, you will no longer have the demonstration program that came with the board. Never fear—you can download the original demonstration code on the Adafruit Learning System page here: *https://learn.adafruit.com/ adafruit-circuit-playground-express/downloads*.

When you save the program—for example, my two animations program—the file saved will be named two-animations.uf2. Make-Code replaces spaces with the dash character -. And the name is suffixed with the .uf2 file extension, as noted earlier, which informs the software on a computer and the firmware on Circuit Playground Express that the contents are to be executed as code. A UF2 file cannot be edited with a text editor on the computer; it can only be modified within MakeCode.

UNDER THE HOOD: JAVASCRIPT

JavaScript is a programming language, commonly used to pro-grammatically build web pages. It turns out that Microsoft is very clever in its construction of MakeCode. While we have been using the blocks in the MakeCode editor, behind the curtain, it has also been writing a JavaScript program.

For most applications, you can forget that the JavaScript is there. You don't need to interact with the JavaScript if you do not wish to. But it's fun to see what is happening.

On the MakeCode screen, look at the top middle. You'll see an oval, one side labeled BLOCKS and the other labeled { } JAVASCRIPT (see Figure 3-26).

FIGURE 3-26. The BLOCKS and JAVASCRIPT buttons at the top of the MakeCode screen

If you click the word JAVASCRIPT in the top bar, the code window will change from MakeCode blocks to JavaScript (see Figure 3-27). If you look at how the JavaScript is written, you can see the loops method to create the forever block and the pause blocks. The light.showAnimation function calls built-in code that creates the animations we selected on the Circuit Playground Express NeoPixels.

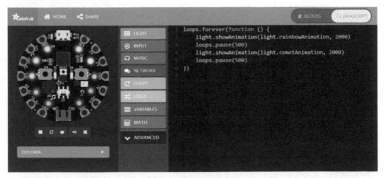

FIGURE 3-27. Clicking the JAVASCRIPT button replaces the Make-Code blocks with JavaScript code. The program behaves the same as the blocks used previously.

JavaScript is not terribly complicated if you compare the MakeCode blocks to the JavaScript. The hard part is knowing which commands represent which actions, and the *syntax* (the way the code must be spelled out) is more complicated and exacting. If you misplace a parenthesis, the JavaScript program fails to compile. For this reason, JavaScript is best left to more advanced projects or those who feel comfortable with learning a more complex programming language. Feel free to click the BLOCKS button to replace the JavaScript with the code blocks. There is little chance you can "break the code" (make it not compile and work) using the blocks in MakeCode.

If you would like to learn more about MakeCode and JavaScript, Microsoft has documentation at *https://makecode.com/docs*.

Your Turn: JavaScript

Change the code in MakeCode, then switch the view to JavaScript. Can you determine what changed? If you adjust numbers in

JavaScript, such as the times of the animations or pauses, does the corresponding MakeCode change also?

WRAP-UP

In this chapter, you have learned to use Circuit Playground Express on a computer and how to use MakeCode to code, download, and execute programs. The next chapter explores more capabilities of Circuit Playground Express using MakeCode.

CHAPTER QUESTIONS

1. Can you directly use either an Apple USB-C or Lightning cable to connect a computer to Circuit Playground Express?

2. What is the address of the website to go to the MakeCode editor for Circuit Playground Express?

3. Can MakeCode be used to program other types of learning electronics?

4. What is the name of the block in which other blocks are placed to have Circuit Playground Express run the commands over and over?

4

Microsoft MakeCode and Interactivity

Microsoft MakeCode is an easy way to access, visualize, and program the Circuit Playground Express. There are MakeCode commands to access nearly all the capabilities on the board. In this chapter, you'll use MakeCode to explore Circuit Playground Express features. I'll introduce some additional methods for creating programs as we need them.

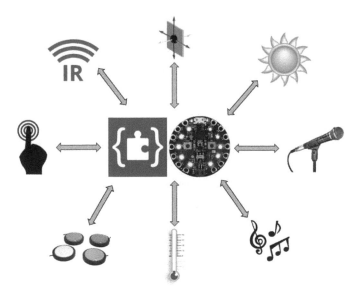

USING BUTTONS

The most influential input device in the history of electronics is the button. Pressing a push button informs an electrical circuit that the user wants something to happen. Circuit Playground Express has two push buttons on its board (labeled A and B) that users can program to do what they wish (Figure 4-1). There is a third, smaller push button labeled Reset between the two larger buttons—it cannot be programmed to do anything by the user since the system uses it to reset the board.

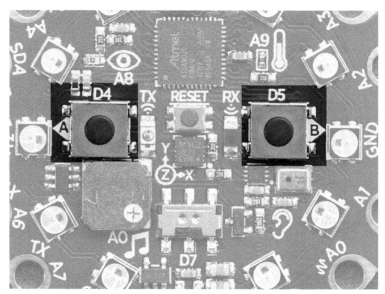

FIGURE 4-1. The two buttons on Circuit Playground Express, "A" on the left, "B" on the right

The push buttons provide interactivity with Circuit Playground Express projects. Code is used to "listen" whether the buttons are being pushed and to do something when activated. Without code, users can press the buttons all they want and nothing will happen, since there is no code "listening" or waiting to see if buttons have been pressed.

The NeoPixel animations in the previous chapter were pre-selected in the code. We could not change the animations after the program started running without changing the code. Buttons allow interactivity while the code is running—for example, "If I press button A, then change the animation."

The code concept that allows Circuit Playground Express to know something has changed while the code is running is called a *variable*. The variable is a number that we'll use in a program. A program may have one variable or many, but code usually does not need many variables to do its job.

What Is a Variable?

A *variable* is a small piece of computer memory that contains a value that might change. Think of a variable as a scratchpad for numbers or information you are working with.

For example, in the previous chapter we used the `pause` block several times, set for 500ms each time. Suppose we wanted to change that to 750ms. We could go through the code and change each block, but if you're using a lot of `pause` blocks, that approach can get tedious.

We could instead create a variable (call it `x`) that contains the value `500 milliseconds` and use it throughout our program in every `pause` block. If we then want to increase the pause to 750ms, we don't need to change every `pause` block. We only need to change the value of `x`!

MakeCode is very clever, so variables are not needed often. In Python and other languages, coders must anticipate when and where they will need variables to read and write information. It's not difficult if we ease into how and why to use a variable when needed.

For the next program, a variable will be used to remember if a button has been pressed. At the start of the program, our scratch-pad variable will be set to 0. Every time the button is pushed, the variable is *incremented* (increased) by 1.

To get started, use your computer browser to open MakeCode at *https://makecode.adafruit.com/#editor*. With a clean programming area, start pulling blocks to make a program that changes animations according to button pushes. From the green LOOPS group, get the forever block (if it is not already on screen) and the new on start block. From the purple INPUT group, drag the on button block to the programming area (see Figure 4-2).

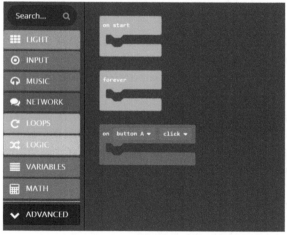

FIGURE 4-2. The main code blocks used in working with buttons

Now click the light blue LOGIC group button and drag the if...then...else block (the one shaped like a letter "E") into the forever block (Figure 4-3).

The if...then...else block is placed inside the forever loop (Figure 4-4).

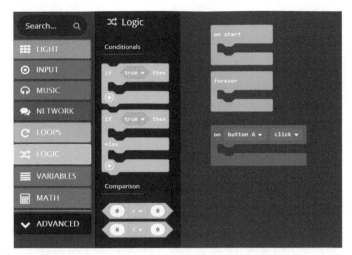

FIGURE 4-3. The LOGIC group of blocks contains the if...then
and if...then...else blocks.

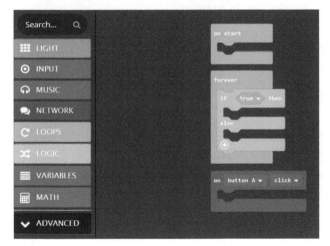

FIGURE 4-4. The if...then...else block added to the
forever loop

The if...then...else block allows a program to ask a question.
In this program, the statement will ask, "What value is the variable we'll create?"

What Are Conditional, *if...then*, and *if...then...else* Blocks?

In code, when a value needs to be compared with another value, an `if` statement makes the comparison. For example, check out the following:

```
If I feel hungry,
    then I will eat an apple,
else
    I will go for a bike ride.
```

This is called a *conditional statement*. It asks what your *condition* is. You will be either hungry or not hungry. If you are hungry, you eat. If not, you ride your bike. Deciding what to do depending on certain conditions is what coding is all about. For a button, this can be written in sentence form as

```
If the button is pressed
    then light the green light.
```

or

```
If the button is pressed
    then light the green light
else
    light the red light.
```

The second example adds this part of the condition: "What if the first part is not true?"

In code, this may look like the following pseudo (not quite) code:

```
if button pressed
    turn green LED light on
else
    turn red LED light on
```

The previous code would have a red LED light always on *unless* the button is pushed, in which case the green LED light would go on. You can also read the word `else` as otherwise.

The `else` part of a conditional may have multiple steps, as many as we want. Here's an example:

```
if food is red
    Your meal is an apple
else if food is orange
    Your meal is a pumpkin
else if food is green
    Your meal is grapes
else if food is yellow
    Your meal is a banana
else
    I don't know what food you are looking at
```

The final `else` is the condition that if "you questioned something multiple times, this is what to do if none of the conditions are true."

The code from this point forward will use the `if` statement block to compare something and make a decision based on whether a condition is true or not (false).

At this point, on the `if...then...else` block, click the circled plus sign (+) five times so we get additional blank `else if` spaces (Figure 4-5). The program uses one `if` statement for each type of animation the program will cycle through.

This action will expand the number of checks, as shown in Figure 4-6.

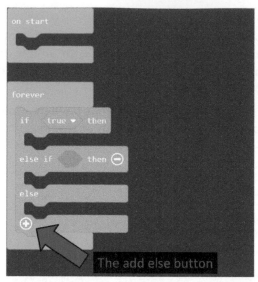

FIGURE 4-5. Expanding an `if` statement with additional `else` cases

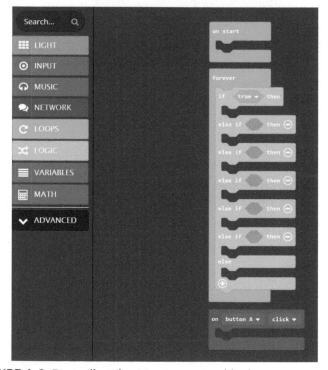

FIGURE 4-6. Expanding the `if..then..else` block to accommodate seven statements

At this point, none of the `if` statements have a comparison to test. The LOGIC group has a `0 = 0` block that is the right shape to put into the `if` statement (Figure 4-7).

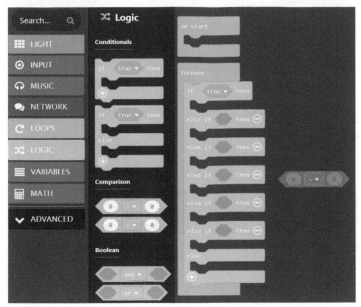

FIGURE 4-7. Selecting the conditional block `0 = 0` from the LOGIC block group

Add `0 = 0` blocks to all the blank slots in the `if` statement.

Now open the VARIABLES block group. If you only see "Make a Variable…" then click that and create a variable named `item`. There are three types of variable blocks, and the program will use all three. Start by dragging the oval `item` block and place it on the left circle on each `0 = 0` conditional we just placed in the `if` statements (`item` replaces the leftmost 0). Each conditional should then read `item = 0` (Figure 4-8).

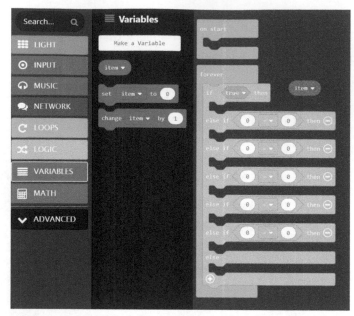

FIGURE 4-8. All of the conditionals are in the if statements.

Place a set item to 0 block in the on start block, as shown in Figure 4-9.

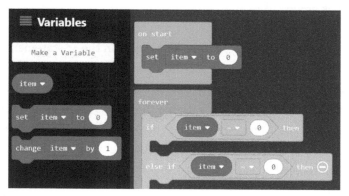

FIGURE 4-9. An item variable is added to each conditional and a set item to 0 is placed in the on start block.

The program next uses the final variable type of block, change item by 1. Add that block by placing it inside the on button A click block (Figure 4-10).

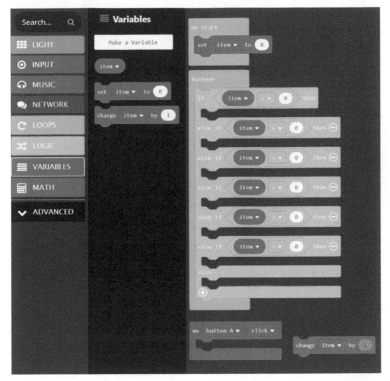

FIGURE 4-10. Adding change item by 1 **to the** on button A click **block**

For each item = 0, change the numbers so they look like the ones in Figure 4-11. Also add a set item to 0 block in the very last slot of the if statement.

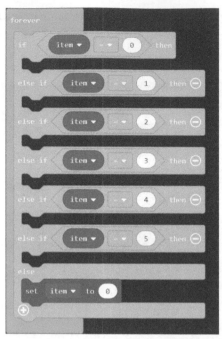

FIGURE 4-11. Setting each if statement to check for a different number, and adding the set item to 0 block at the end

The next step is to fill in all the empty spaces in the if block. Open the LIGHT block group and put a show animation in each of the vacant block spots (Figure 4-12).

The Positioning of Code Blocks on the Screen

The blocks that are rounded at the top, shaped like the letter C, such as forever and on start, do not have to be stacked in a straight line up and down, as shown in the examples. They can be anywhere in the programming part of the screen. You are free to position them to make your code readable to you and others.

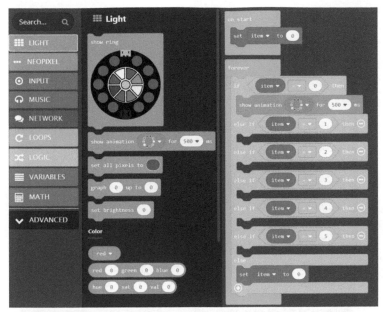

FIGURE 4-12. Adding show animation blocks to the if statement

Once you have all the show animation blocks filled in, the adding blocks part is done. At this point, change the second through sixth show animation blocks to show one of each type of animation, as in Figure 4-13.

The completed program is shown in Figure 4-14.

Go ahead and try running the code in the simulator and on a physical Circuit Playground Express. Pressing the A button changes the animation, cycling through all six light sequences before returning to the first.

Ensure you save a copy of your program at this point. We will change it later and we must be able to get back to that program at some point.

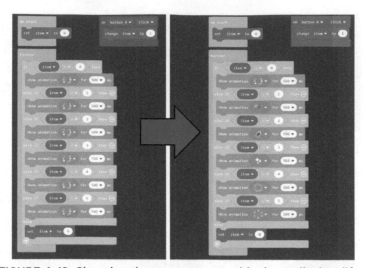

FIGURE 4-13. Changing the show animation blocks to display different animations

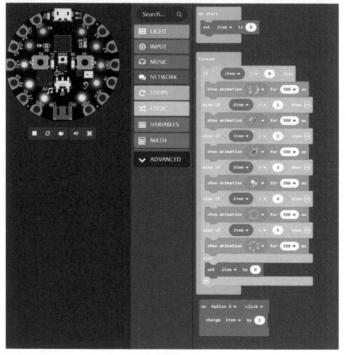

FIGURE 4-14. The final push button program. It's a bit more work than previous examples, but worth it.

How does it work? The on start block sets the variable item to 0. It is clearer now that the value of item is the type of animation the board should display. The on button A click block increments the variable item by 1. The variable item then tells the forever loop which animation the user wants to display. There are six animations in MakeCode. If the button is pressed and item is incremented from the number 5 to the number 6 (higher than the number of animations available), the last else statement sets item back to 0, indicating the first animation.

This code would work well in a wearable project if you want to try different animations of the NeoPixel lights. Or it could be used as an accent light, perhaps under a shelf, lighting up the area below.

Your Turn: Buttons

The challenge: Change the code to use the B button to go backward through the animation list (very handy if you skip over one animation and don't want to click the button six more times).

Hint: You can have more than one on button click block. You have the option to change A to B.

SHAKE, RATTLE, AND ROLL

A very satisfying way to make interactivity, especially in wearable electronics, is to have a project react when it detects movement. The most common reaction is to have the project blink a light, although making various sounds or emulating a computer mouse are also popular.

The hardware on Circuit Playground Express that provides motion sensing is the LIS3DH 3-axis XYZ accelerometer chip (Figure 4-15), located in the center of the board. You can use it to detect tilt, gravity, and motion, as well as tap and double-tap strikes on the board.

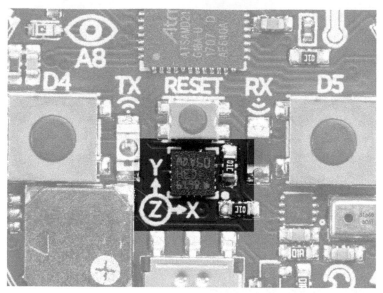

FIGURE 4-15. The accelerometer motion-sensing chip on Circuit Playground Express

You can see that there is marking next to the chip on the board. The mark indicates that when you have Circuit Playground Express on a flat surface with the USB port at the 12 o'clock position, that movements straight ahead are positive Y, movements to the right are positive X, and movements bringing the board straight up off the surface are positive Z.

What does all this mean? The sensor measures changes in movement in three directions—the three dimensions of the space around us. Circuit Playground Express can measure this movement and assign number values to each of the three ways in which the board moves.

Let's return to the MakeCode display in your computer web browser. If the MakeCode screen is not already in front of you, go ahead and set it up as discussed in Chapter 3. Click New Program to start fresh.

Click the `forever` block and press Delete on the keyboard to remove it. Do the same for any other blocks until the screen is clear. Go to the purple INPUT block and drag the `on shake` block to the code part of the screen. Your screen should look like Figure 4-16.

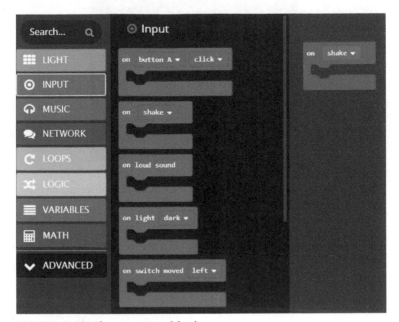

FIGURE 4-16. The `on shake` block

The new `on shake` block watches the accelerometer chip. If it registers any significant values, which indicate movement, it will execute the code inside the block. It reads the three X, Y, and Z values and performs all the comparisons in the background. Very handy!

Notice the small purple SHAKE button on the Circuit Playground Express simulation picture has appeared (Figure 4-17).

This button is shown when you add code blocks that use the accelerometer.

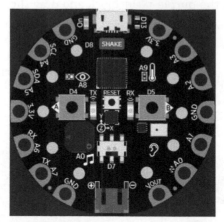

FIGURE 4-17. The colored SHAKE button on the Circuit Playground Express simulator between the black microcontroller square and the silver USB socket

Move your mouse around the picture of the board or click the SHAKE icon within the simulator with your mouse. This makes the board look like it is bending and twisting. The MakeCode environment does that to simulate shaking the board as if it were in your hand.

Next, add the code for making lights come on when the board is shaken. This will require two blocks from the light blue LIGHT block group: show animation and clear. clear is near the bottom of the list of LIGHT blocks.

Now change the value 500 in the show animation block to 2000, as shown in Figure 4-18. This change allows the animation to go on for two seconds after a shake instead of the default 500ms (half a second).

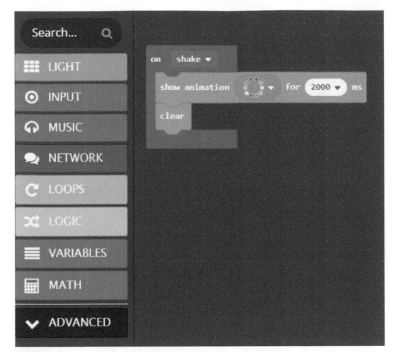

FIGURE 4-18. The shake code

The code is done. Click the purple SHAKE button on the Circuit Playground Express simulator. You should see the board light up for two seconds when shaken.

You can now download the code to Circuit Playground Express to see it for yourself.

Your Turn: Shake (Part One)

Change the animation to one of the other five animation types.

MAKING THE ACCELEROMETER DISPLAY MULTIPLE ANIMATIONS

Looking at what numbers the accelerometer outputs to the microcontroller is fun in the scientific sense. A very popular use for

Circuit Playground Express is to use an accelerometer shake as a method of changing the color of the NeoPixels. The code blocks from the previous chapter can be added to the shake code we made in this chapter.

Build the code shown in Figure 4-19 by dragging the appropriate color blocks into the program area. Do not worry about any values for the moment; focus only on the blocks and what order to place them in.

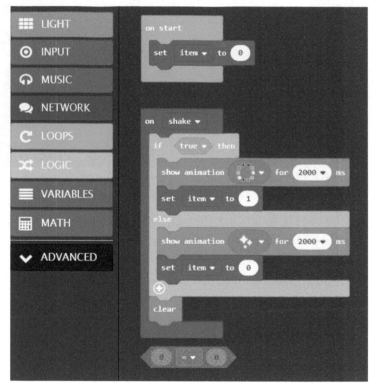

FIGURE 4-19. The code for changing lights as a result of shaking

The set item to red VARIABLES block creates a number in our scratch memory. The program uses one value, item, to remember what type of animation we are currently using so we can change to another one.

Change the default values of certain blocks:

1. Change the animations to run for 2000ms (two seconds) each.

2. For the second set item to block (the first in the on shake loop), change the value from 0 to 1.

3. For the comparison, go to the red VARIABLES group to get the a variable named item and place it in the left side of the 0 = 0 block in place of the zero.

4. Drag the comparison block to replace the value true in the if statement.

Your code should look like the program in Figure 4-20.

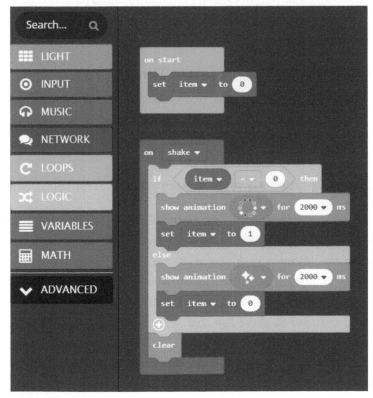

FIGURE 4-20. The working shake and lights code

It's a bit complex, but the code does exactly what we want. Try it in the simulator. If you shake Circuit Playground Express, it shows the rainbow animation. Shake it again, and it shows shimmering white lights. Shake once more, you get the Rainbow animation again, and so on.

If you have your hardware Circuit Playground Express, go ahead and download the code. MakeCode will create a UF2 program file, which you will put on the flash memory on Circuit Playground Express (see Chapter 2).

Note that one shake makes rainbow lights and another shake makes shimmery lights. This code, as is, is very often used in wearables. Makers have sewn their Circuit Playground Express into badges, skirts, shoes, and even hair clips and hats. Interactive lights are very popular projects.

Save your code at this point.

Your Turn: Shake (Part Two)

The animation stops after two seconds and the lights go out. Change the code so that more animations will run and so that the lights will remain on at the end of the animation.

USING THE SLIDE SWITCH

If you are using Circuit Playground Express in a blinky project or on a piece of clothing, you may want to change the animations according to your choice. For example, the project displays some animation like rainbows until your mood changes, and you decide you would like shimmery lights instead. You could use the push buttons to make the change, but let's use another Circuit Playground Express feature: the slide switch. This component is located between the center accelerometer and the battery plug (see Figure 4-21).

FIGURE 4-21. The location of the slide switch

It would be perfect to have one animation display if the switch is moved to the right and another animation display if the switch is moved to the left. Then your animation remains the same until you change it.

Figure 4-22 shows the MakeCode for changing an animation based on the position of the slide switch.

Most of the MakeCode is reusing blocks previously used:

* The `on start` block sets our one variable `item` to 0. This variable helps determine if the switch is moved.

* The `forever` loop checks to see which position the switch is in by checking the variable `item` using an `if` statement block. If `item` is 0, then the switch is in the position it was when the program started. If the variable is 1, then the switch has been moved in the opposite direction.

* One new block is used twice: on switch moved from the pink INPUT group. Place two copies of this block into the code space. One should be set to on switch moved right, the other to on switch moved left.

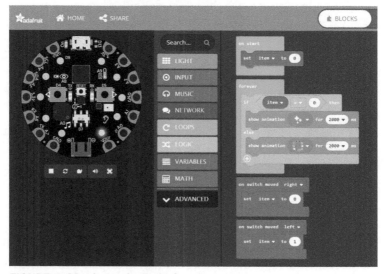

FIGURE 4-22. The MakeCode for changing an animation based on the slide switch's position

Complete the program and try it in the simulator. Double-check that the slide switch does change the animation. If not, check your program again against the sample shown. If the simulation is working, go ahead and load the code to Circuit Playground Express. With this code, your project will be selectable between two different animations. If this project is used as a decoration or on a costume, the animation is changeable via a simple switch.

Figure 4-23 shows one final MakeCode program. The program combines the concepts in the chapter to provide a very classy effect. The code will blink an animation on the NeoPixels when shaken. If you wish to change the animation, you move the slide switch. So your Circuit Playground Express is, say, rainbow when you move. Then you can change it later with the switch to be blinky instead. This code saves you from having to bring a computer and change the code when you tire of the animations.

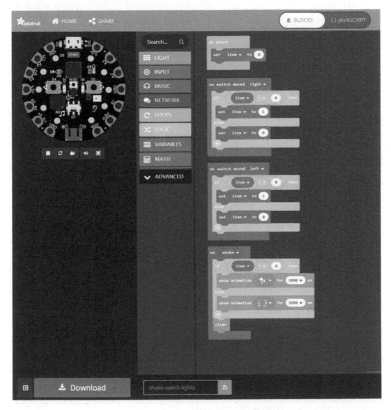

FIGURE 4-23. Animations when Circuit Playground Express shakes. This version uses the slide switch to change animations.

Most of this code should seem familiar from previous examples. The code would be much simpler if there were a condition to check for "Has the slide switch changed?" but there is no block for that function. The variable item is used to hold the fact that the switch has changed.

The animation does not run continuously in this example. It runs for two seconds after a shake is detected, and then stops. The clear block at the end of the on shake loop ensures the NeoPixels are off. For wearables, this saves battery life.

To recap, this code will display colored lights when shaken, and it will sparkle white lights if the slide switch is moved to the other position. Be it a snow globe or a colorful wearable, this code is very useful.

Once more, save your code at this point.

YOUR TURN: SLIDE SWITCH

The challenge: Load or code the simple if shake then set animation code from the start of the "Shake, Rattle, and Roll" section. Add code blocks that will allow the code to work if the slide switch is in one position but disables the animation if the switch is in the opposite position.

How might the code for this challenge be useful? The slide switch can be used to "turn off" what Circuit Playground Express is doing to save power or stop making lights on movement. It is not a true off/on switch, since it can be programmed via code to do many things. But if the slide switch is not in use for other input, having the ability to stop what certain code is doing (rather than unplugging the power) is very handy and it does not take much code to achieve that result.

JAVASCRIPT

For JavaScript fans, each of the programs written for this chapter have their corresponding JavaScript code. The corresponding JavaScript is accessible by clicking the JAVASCRIPT button next to the BLOCKS button in the top center of the MakeCode screen in the web browser (Figure 4-24).

FIGURE 4-24. To switch from MakeCode Block View Mode to JavaScript Mode, use your mouse to click the JAVASCRIPT button. Clicking the BLOCKS button changes the view back.

The JavaScript for the slide switch code is shown in Figure 4-25.

```javascript
let item = 0
input.onSwitchMoved(SwitchDirection.Right, function () {
    item = 0
})
input.onSwitchMoved(SwitchDirection.Left, function () {
    item = 1
})
item = 0
loops.forever(function () {
    if (item == 0) {
        light.showAnimation(light.sparkleAnimation, 2000)
    } else {
        light.showAnimation(light.rainbowAnimation, 2000)
    }
})
```

FIGURE 4-25. The JavaScript for the slide switch changing animations code written in Figure 4-23

The JavaScript code turns out to be very readable. This is due to precoded functions in JavaScript that handle the Circuit Playground Express specialized hardware and capabilities. Methods are available for checking the switches and displaying NeoPixel animations.

JavaScript can be used instead on the block editor, but learning JavaScript is more than can be accomplished in this Getting Started book.

WRAP-UP

We covered use of the various switches on Circuit Playground Express. We also reviewed the accelerometer and how to use it in MakeCode when the board is shaken or tapped.

You also learned about new code blocks in MakeCode:

* Variables

* if .. then conditional checks

* The on start block used to set up the board prior to the forever code running

* New blocks such as on button, on switch, and on shake that react to something happening

Behind the scenes in the web browser, MakeCode is also building JavaScript code that is identical to the MakeCode block program. The programmer should not need to change the JavaScript, because the MakeCode block code is identical to the JavaScript code in functionality.

CHAPTER QUESTIONS

1. Besides the on start and forever code blocks, can you list other code blocks that can also host other code blocks?

2. Can the Reset switch be used in normal program interactivity on Circuit Playground Express?

3. Can the on shake block contain other code blocks? If so, name some.

4. Can Circuit Playground Express be programmed in JavaScript without using the blocks editor?

5

. . . .

Advanced Microsoft MakeCode

In this chapter we are going to use Microsoft MakeCode to take advantage of more of the Circuit Playground Express components and combine them for interesting effects.

If you have not browsed all of the block groups in MakeCode, I encourage you to do so. There are 15 code block groups to explore. Actually, the EXTENSIONS block group contains even more groups—and the possibility for more as the product evolves.

With the ease of MakeCode, it takes only a couple of minutes to get something interesting, put it together, and see what happens. The projects in this chapter will provide some ideas on a variety of applications.

SOUND AND MUSIC

Circuit Playground Express uses hardware (Figure 5-1) to convert digital signals to analog output to create sound via the onboard speaker. This is in contrast to many other microcontroller projects that toggle a digital signal to a piezo transducer, which is harsh sounding. While the sound from the Circuit Playground Express is good, the speaker is still *monophonic*, outputting to one channel as opposed to two (stereo).

MakeCode has built-in code to take advantage of the Circuit Playground sound output in an easy-to-use format.

FIGURE 5-1. The Circuit Playground Express speaker (right) and amplifier (left)

What Is Sound?

Sounds are vibrations in the air around us. Sound can be generated by movement—our hands clapping, a bird chirping, or a musical instrument. Our ears are sensors that translate the air pressure changes into electrical signals, which our brains then interpret. Mechanical systems can do the same translation that our ears do for electronics; these are called *microphones* or *transducers*.

One aspect of a sound is its *frequency*. Frequency refers to how fast the air pressure changes when a sound is made. Think of a big tuba: the sound is a low bump, bump, bump. Then a flute: a high-pitched tweet, tweet. Those sounds are at different sound frequencies (see Figure 5-2).

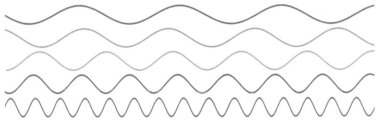

FIGURE 5-2. Representations of sounds: the red at a low frequency, purple at a high frequency. Source: Wikipedia

Frequency is measured in cycles per second, called *hertz*. The sound in the red line above has fewer cycles during the period of time represented, so the frequency is less than that of the sound representations below it. The purple line has the highest frequency. If we connect each wave to a musical instrument, the red one could be a tuba and the purple one a flute, with other instruments between them.

A sound may, by design, be a short alert or a longer melody. Emergency tones are usually sounded for a minute or more, whereas a piano key is most often struck once and the sound is quite short. That is *duration*, the time a tone is heard that we record in units of seconds.

Finally, the height of each frequency represents how loud a sound is. Loudness, or *volume,* is a measurement of how loud the sound is to our ear. You can adjust a music player to play very softly or at ear-splitting loudness.

Music

Knowing what sounds are and how we characterize their properties, we can better appreciate what we call *music*. Music can basically be defined as "sound organized in time." Most of us think of music as not just sound, but specific musical notes. What are musical notes, and how can we get Circuit Playground Express to play them?

Each musical note represents a specific frequency and duration of sound. Western musical tradition is built around seven main notes (and five additional "accidental" notes). You probably know them as "Do-Re-Me-Fa-So-La-Ti." You can sing those notes in a low *pitch* (at low frequencies) or a very high pitch (at higher frequencies). In musical notation, the syllables above are given the letters C-D-E-F-G-A-B. In some languages other symbols are used, but the English convention is widespread.

MakeCode has code blocks for easily playing musical notes on the Circuit Playground Express speaker. The play tone block will output a tone at a selected frequency for a predetermined length of time, called a *beat*.

Start a new MakeCode project and select the blocks for the program in Figure 5-3. The musical blocks are in the bright red MUSIC block group.

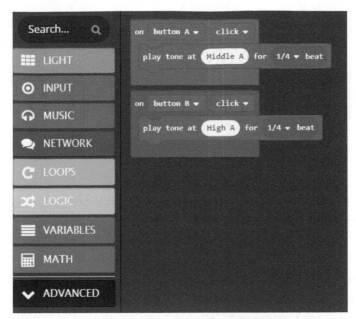

FIGURE 5-3. A very simple piano playing two different musical notes

You can use the simulator in MakeCode to press the Circuit Playground Express Button A and Button B. What happens? Button A plays a tone when pressed and Button B plays a tone as well. In the program, the tones are both named "A" but one is a Middle A and the other a High A. When selecting the value to put into the play tone block, you'll see a piano-style keyboard appear (Figure 5-4). A number is in the circle, and when your mouse hovers over a key, the key name is displayed below the keyboard. When you select the piano key marked "Middle A," the number in the play tone circle is 440.

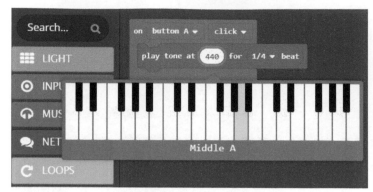

FIGURE 5-4. Clicking the frequency value circle brings up a keyboard, if needed.

It turns out the musical note called "Middle A" by musicians is called "440 hertz" by those who measure frequency. Each musical note has a specific frequency. For the Button B block, when you hover your mouse over "High A," you get the value 880 (hertz) and a key more to the right of the keyboard.

You can select different musical notes in your code and then play them on Circuit Playground Express. As you change the values, you hear different musical tones. Each separate tone is called a *note*; 440 hertz is the frequency for the musical note called "Middle A." Doubling the frequency from 440 to 880 hertz makes the note sounds different; 880 hertz is called a "High A." Mathematical relationships exist between musical notes, stemming from how human ears hear the notes that make up music. If you want to see a chart of the frequency for each musical note, visit *www.arduino.cc/en/Tutorial/toneMelody*.

How can MakeCode output music with multiple notes, to form a *tune* (a sequence of notes)? Basic music is made by selecting different notes for a certain length of time. The MakeCode in Figure 5-5 allows Circuit Playground Express to make a part of a tune. Note the values for the beat (length of tone) for each musical block to make the tune sound right.

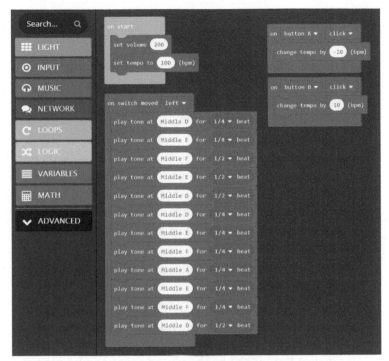

FIGURE 5-5. MakeCode that plays a bit of a tune

In the simulator (or if you've uploaded the code to your Circuit Playground Express, on the board itself), move the slide switch to the left to trigger the tune. If the switch is already at the left, move it to the right, then back to the left again. The tune will play. Move the switch to the right, then to the left again, and the tune will play again.

You can see that playing even a partial tune takes a fair number of blocks. Your favorite song would be much, much longer to set up in block code. Each note has a specific duration (time the note is played). Some songs add *rest* (pauses between notes). If these values are changed to something random, you start to get a musical mess. Music turns out to be a very structured list of sounds.

Is the Board Polyphonic?

With only one speaker and limited hardware capability within the Circuit Playground Express microcontroller, the sound output is limited to making one musical tone at a time (*monophonic*).

Polyphonic is the ability to play more than one musical note at a time, similar to the capability of a piano or a musical synthesizer. Many forms of music require the ability to play multiple notes at once for the piece to sound and play correctly. Unfortunately, Circuit Playground Express does not have the capability to do multiple-tone output.

Songs with simple notes played *sequentially* (one at a time) work fine.

Circuit Playground Express can also play short clips of sound (often called WAV or wave files) that may contain polyphonic sound. But limited hardware capability makes the length of the sounds that can be stored and played rather short. Playing WAV files in Circuit-Python is discussed in Chapter 7.

This program also allows the *tempo* to change. Tempo is the speed or the pace of the music. Normally, if you want to speed up or slow down a piece of music, you need to manually change the durations of the notes. MakeCode has a feature that allows tempo changes behind the scenes. You can press the A button or the B button to slow down or speed up the tune in this program.

How Can Sound Help My Project?

Projects up to this point have used light to provide indications. Lights work well if you are constantly watching them, but if you look away, you may miss the information the lights want to display.

With sound, projects may be created that are silent until an alert is needed. Here are some of the uses for this:

* Make a sound if light is detected

* Make a sound if something is too hot or too cold

* Make a sound if movement is detected

LISTENING FOR SOUNDS

Circuit Playground Express also has a microphone for sound input. The microphone is a digital type (see Figure 5-6) that may be different from analog microphones used in other microcontroller projects. This section will demonstrate how to use the Circuit Playground Express microphone in MakeCode.

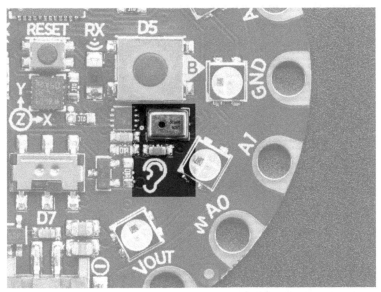

FIGURE 5-6. The Circuit Playground Express microphone

The firmware for the microphone is still under development as of this writing, so the number of functions available is not as expansive as for other sensors. The engineers at Microsoft have used the microphone in MakeCode for a very useful feature: loud sound detection. Detecting a change in sound has many uses in everyday life. You can have Circuit Playground Express listen for a noise such as a door opening or closing. It could listen for the loud sound made by a car horn or a washing machine/dryer buzzing at the end of a cycle.

The next project uses sound detection to mimic a product that performs an action if two hand claps are made within a short time. The commercial product turns on and off electrical appliances like a light. With the Circuit Playground Express version, you are not limited to performing only one action. You can turn NeoPixels on and off to simulate a lamp, or you can program other actions such as playing a song, simulating a computer keyboard, or any other actions you can dream up.

The on loud sound main block is in the purple INPUT block group. Drag it onto a blank MakeCode screen (Figure 5-7).

FIGURE 5-7. The on loud sound main block in MakeCode

This block will execute the code within it if you make a loud sound such as a hand clap. You can try the code in Figure 5-8 to have the NeoPixel lights change every time one clap is heard by Circuit Playground Express.

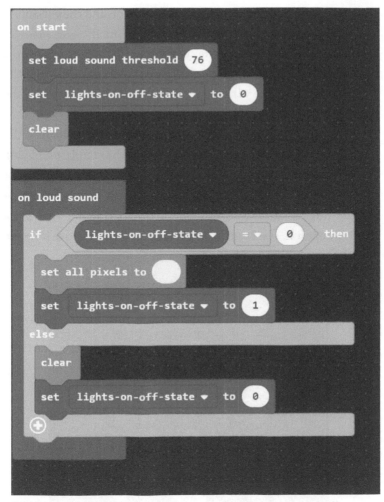

```
on start
    set loud sound threshold  76
    set  lights-on-off-state ▼  to  0
    clear

on loud sound
    if  ⟨ lights-on-off-state ▼  = ▼  0 ⟩  then
        set all pixels to  ⬤
        set  lights-on-off-state ▼  to  1
    else
        clear
        set  lights-on-off-state ▼  to  0
    ⊕
```

FIGURE 5-8. A single loud noise turns the lights on and off.

In this code, a threshold (an upper limit) to the loudness of the sound can be set. A variable called lights-on-off-state is created and set to 0 to track whether the lights are on or off, and to ensure that the NeoPixels are off. Then the code waits in the on loud sound loop for a clap. When the board "hears" a loud noise, it executes the code. In our example, if the variable lights-on-off-state is 0, the lights must be off, so they are set to on and the variable is set to the value 1 to help remember they are on. If `lights-on-off-state` has the value 1, then the code knows that there has been a previous clap and the new clap will turn the lights off using the `clear` block code. The variable is set to 0 to let the program know the lights are off.

Try the code yourself on the simulator or by uploading the code to your Circuit Playground Express. In the simulator, having an `on clap` block triggers a circle to appear to the bottom left of the picture of the board (Figure 5-9). Using the mouse, move the line within the circle to raise and lower a virtual sound level. For a clap simulation, start with a low value and quickly move it to the top for a high level and back down again.

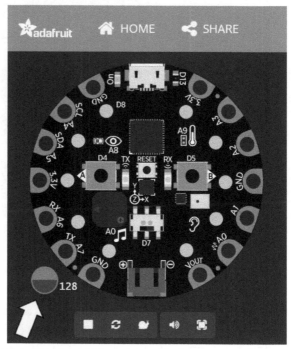

FIGURE 5-9. The sound level simulation circle appears when an on sound block is used. Grab the horizontal line in the circle with a mouse and move it down, then up, then down to simulate a clap.

What will happen in this program if you clap twice?

The first clap will turn the lights on; then the second will turn them off. This is not exactly like a commercial clap detector. Try another loud sound like a cough. The board will most likely think that sound is a clap and toggle the lights. To prevent unintentional loud sounds from activating the lights, the commercial product looks for two loud noises within a short duration. It is less likely that two loud sounds might occur within a second or two, so it makes for a good signal to activate a device.

The previous code needs to be modified to tell if two claps happen within a certain time. It turns out that 1.5 seconds is a good value, and this is what the code in Figure 5-10 will test for. The value may be changed later as needed.

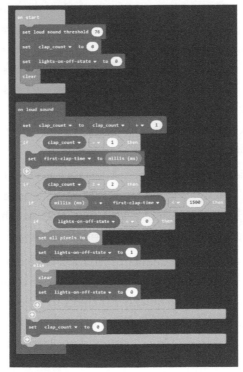

FIGURE 5-10. The final two-clap activation code

For this version, another variable is needed to count claps. Create a new variable called clap_count.

When the loud noise is detected, the clap_count variable is incremented by 1 to note a clap was detected. Then different actions are done depending on the number of claps:

* If the number of claps counted is 1, then the current time Circuit Playground Express has been running is stored in the variable first-clap-time. While the board does not have a clock telling the time of day, it can keep track of the time since you plugged it into power, which is also helpful. That time is available in your programs via the green block millis (ms), which displays the board running time in milliseconds (thousandths of a second).

* If the number of claps detected is 2, then the fun begins. The current time from the millis (ms) block is read and the previous time stored in first-clap-time is subtracted. This results in the time between the first clap and the second clap. If that time is less than or equal to my chosen 1500 milliseconds (which is the same as 1.5 seconds), then it goes to the code that toggles the lights as used in the previous example.

* If, for some reason, the board detects two claps but the claps are not within 1.5 seconds, then clap_count is set back to 0. This helps if you were to clap once, then cough perhaps 2 seconds later—that would not trigger the board.

Try this code in the simulator. You will have to move the line in the sound circle up and down quickly twice to get the lights to turn on, then twice quickly for them to go out. Upload the code to the Circuit Playground Express board. Now you can try the clapping yourself. One clap should not trigger it, and two or more claps spaced more than a second and a half would likewise be ignored. Two quick claps will set it on; two more quick claps will set it off. You can change the value 1500 to, say, 1000 for a fast activation or 2000 for a slower activation.

Be sure you save your code for reference later.

Your Turn: Sound Sensor

Change the code so that

* two claps plays some musical notes

* it takes three fast claps to trigger

Project Ideas: Advanced Clapping

Here are a couple of ideas for taking the sound activation concept to more advanced projects.

* MakeCode allows for external NeoPixel control—that is, hooking a strip of NeoPixels up to one or more of the data lines and controlling them. A subblock of the LIGHTS group, named NEOPIXEL, controls external NeoPixels connected to Circuit Playground Express. See Adafruit product ID 3812 for external NeoPixels.

* Here's an advanced technique: Lamp control, like the commercial Clapper, is possible with certain components. Adafruit sells the Controllable Four Outlet Power Relay (*www.adafruit.com/product/2935*). One control pin and ground can be connected to the power relay green terminals and turned on and off via MakeCode blocks in the red PINS block group. The power relay safely isolates the small signals used on Circuit Playground Express and the dangerous power line voltage and current. Remember: Never connect any wall outlet wiring directly to Express; it could cause serious injury.

WRAP-UP

Sound is an excellent notification method. Many people focus a project to use multicolored lights to notify the user of a situation. If distracted, a person may not see the lights. Sound is an attention-getter that does not need to be watched to notice.

Sensing sound is an especially good method to alert a user in case a radio communications method like WiFi or Bluetooth is not available. People with limited mobility might be able to clap to have an action done. Sound detection is also useful in alarm systems to signal the breaking of glass or the sound of a door opening. Once you've explored the basics of sound detection, look into additional projects with larger microcontrollers that perform speech recognition (which takes more processing power than Circuit Playground Express has onboard).

CHAPTER QUESTIONS

1. Take a short tune and work to translate it into MakeCode. Feel free to browse the Internet for songs already transcribed by others.

2. Can you describe how Circuit Playground Express might make a room alarm by sensing a sound and alerting a person inside?

6

Coding with CircuitPython

In this chapter, an alternate way to program Circuit Playground Express is demonstrated using a language called *Python*. Python is the fastest-growing programming language in use today and is taught in schools and universities. It's a high-level programming language, which means it's designed to be easy to read, write, and maintain. Figure 6-1 shows Blinka, the CircuitPython mascot.

FIGURE 6-1. The Adafruit CircuitPython mascot: Blinka

The change from using the graphical Microsoft MakeCode to using text-based programming may seem like a huge one. It is not. The Python programming language has been gaining a steady following, significantly in the Raspberry Pi community. Why? Because Python is easy to use, and because the language was designed from the start to be easy to improve, extend, and grow. The same Python code can run on a small wearable, on to other computers, on up to a supercomputer, with perhaps only a slight change.

Python includes modern programming commands and it supports code extensions, called *modules*. (In some computer languages, like Arduino, these are called *libraries*.) Modules are code packages that can be used by a Python program to perform specific tasks. For example, there are modules to perform complex number crunching or to graphically plot data. And nearly all modules are open source software—code available on the Internet at no cost and freely shared.

Module vs. Library

There's some discussion in the computing field as to whether code that is imported into Python is a module or a library. To remain consistent with CircuitPython documentation, I'll use the term *library* most often. Just think of the terms as interchangeable for this book.

Unlike the Arduino environment, where all coding is done on a desktop or laptop, compiled into machine code, and then loaded onto the circuit board, Python is an *interpreted language*. This means that the hardware can interpret and act on each command you type, practically instantaneously. There's no need to compile and upload your code to see if it works.

CircuitPython provides interactivity via a Read–Eval–Print Loop (REPL, pronounced *rep-ul*). On your computer you can type Python commands into the REPL, and the board will process and respond to each line of programming entered. This allows the user to see what specific commends do in real time rather than performing multiple steps to get code into a processor for execution.

CIRCUITPYTHON VS. OTHER PYTHON IMPLEMENTATIONS

CircuitPython is the implementation of Python created by Adafruit for several products, including Circuit Playground Express. It is a *fork* (derivative) of MicroPython, a version of Python written by Damien George to run on microcontrollers.

CircuitPython adds hardware support for a range of Adafruit Industries microcontrollers. It allows users with limited hardware experience to easily program their devices. No previous experience necessary—it's really simple to get started!

All CircuitPython code is run from the Circuit Playground Express internal flash drive/thumb drive, the space used previously to put MakeCode onto the board.

CircuitPython excels at the following:

* **Very fast development:** Write the code, save the file, and it runs immediately. No compiling required.

* **REPL:** You can start interactive programming with the REPL.

* **Easy code changes:** Since your code lives on the flash drive, you can edit it whenever you like, and you can also keep multiple files around for easy experimentation.

* **It's Python!** CircuitPython is completely compatible with Python (it just adds hardware support).

* **Strong hardware support:** There are many more modules for external sensors and capabilities than in MakeCode (but not as many as Arduino, yet).

* **File storage:** CircuitPython's data storage ability makes it great for data logging, playing audio clips, and otherwise interacting with files. MakeCode doesn't currently have support for file storage, and Arduino has limited support at present.

See Appendix B for more Python resources.

INSTALLING CIRCUITPYTHON

If you are sure your board already has the latest CircuitPython release, you can skip this section (for example, a teacher says you are all set or you have already placed CircuitPython on your Circuit Playground Express).

If you would like the latest version of CircuitPython, go ahead and follow along. Updating the software is the same as installing a fresh copy.

How do you know which version of CircuitPython is on your Circuit Playground Express? Connect your Circuit Playground Express to your computer. After a moment, the flash drive CIRCUITPY should appear (Figure 6-2). If your Express does not show a new CIRCUITPY drive but shows a CPLAYBOOT drive, then it needs to have CircuitPython loaded; see the how-to in the next section.

If the board is providing a CIRCUITPY drive, you should see a file on the drive called boot_out.txt; see Figure 6-3.

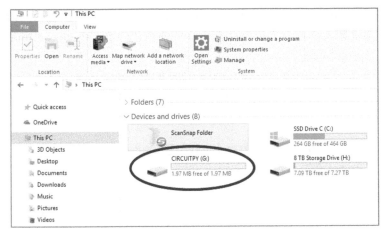

FIGURE 6-2. The CIRCUITPY flash drive in Windows Explorer

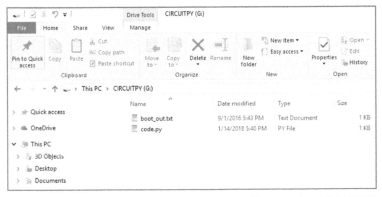

FIGURE 6-3. The `boot_out.txt` and `code.py` files on the CIRCUITPY flash drive

If you open the `boot_out.txt` file (usually by double-clicking it with the mouse), the file contains information on the version of CircuitPython placed on the board. When I updated my board for this book, the `boot_out.txt` file contained the following:

```
Adafruit CircuitPython 3.0.0 on 2018-05-04; Adafruit Cir-
cuit Playground Express with samd21g18
```

You can compare the version number (in this case 3.0) with the latest version on the Adafruit website and decide if a more up-to-date version is available.

In the off chance you believe the board has failed and Circuit-Python no longer works as it did, you can install the latest version to set it up fresh.

Downloading the Latest Version of CircuitPython

When you looked for CircuitPython on your Circuit Playground Express, you might not have found it. If there is no CIRCUITPY drive, there might be a CPLAYBOOT drive. This indicates the board was being used for MakeCode or something else (see Figure 6-4). No worries—this is easily fixed. You can also check to ensure your CircuitPython installation is up to date.

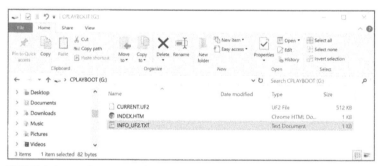

FIGURE 6-4. When you plug Circuit Playground Express in and you get a CPLAYBOOT drive rather than a CIRCUITPY drive, don't worry—that's easily fixed.

Use your Internet web browser to go to *https://github.com/adafruit/circuitpython/releases/latest*. Scroll down to the list of CircuitPython files, and choose the file that contains the text `circuitplayground_express` in the filename. When you click the file, the operating system will display a box that asks you where to save the UF2 file that is the CircuitPython code. You can save it to any file directory you like; just remember where you save it. On Windows this can be the desktop or the Downloads or Documents folder. You can also save the file to a personal flash drive.

> ## Remembering the Gotchas from Chapter 2
>
> If you are running Windows 7, you will need a software driver installed to have your computer recognize the Circuit Playground Express board correctly.
>
> Be sure you use a high-quality USB-to-MicroUSB cable with both power and data lines. Old cables, or cables that are used only to charge another device, will not work and will almost certainly lead you to frustration.

Plug your Circuit Playground Express into your computer and ensure the green power LED is on. Find the Reset button on your board. It's the small button located in the center of the board.

Tap this button once to enter the bootloader. The NeoPixels on the board will flash red and then stay green. A new drive will show up on your computer. The drive will be called CPLAYBOOT (see Figure 6-5).

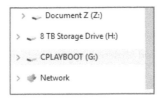

FIGURE 6-5. The CPLAYBOOT drive appears when you reset the board.

If you do not see this drive on your computer, don't be discouraged. Tap the Reset button twice. The rhythm of the taps on the Reset button needs to be correct, and sometimes it takes a try or two. If you have a Circuit Playground Express and it's fresh out of the bag, pressing the button once will probably do it.

The number of presses to get the green circle and CPLAYBOOT drive depends on what software was used previously: MakeCode, CircuitPython, or Arduino. There is no "wrong" way—use one press or two, as long as the CPLAYBOOT drive shows up in the operating system.

Take the UF2 file you saved from the web and drag it onto the CPLAYBOOT drive (or otherwise initiate a copy of the file to the Circuit Playground Express CPLAYBOOT drive). The red D13 LED will flash as the file is transferred and the NeoPixels will blink and then go out.

If everything is successful, your computer will show a new flash drive named CIRCUITPY. This is your indication that you now have CircuitPython ready on your Circuit Playground Express. Your computer, for instance Windows 10, may pop up a message that it is setting up a new device called Circuit Playground. This is fine.

On the drive, you should see a file named boot_out.txt on the CIRCUITPY drive.

Now it's time to check whether you have the latest version of CircuitPython on your Circuit Playground Express. Open the boot_out.txt file, and you will see information on the version of CircuitPython that was placed on the board:

```
Adafruit CircuitPython 3.0.0 on 2018-05-04; Adafruit
CircuitPlayground Express with samd21g18
```

It is possible that your board will show a version number greater than 3.0. Version 3.0.0 was released in 2018 (with additional releases planned after that) as Adafruit provides additional functionality. samd21g18 looks like a gamer tag but it is the manufacturer's name for the microcontroller on the board.

Do I Have to Install CircuitPython Regularly? Must I Upgrade?

You have to install CircuitPython only once unless you are switching back and forth between CircuitPython, MakeCode, and Arduino. Installing once, you are free to code all you like without going through the install process again until you want to upgrade. Upgrades from Adafruit may come out periodically, adding additional features or fixing issues noted by users. Often, it is worth upgrading if you want to use new features or there is a major update. If you have a project you intend to use and it works well, consider keeping things the way they are.

I'm Having Weird Installation Issues!

Don't worry! The problem could be a corrupt flash drive, which is not the end of the world (or your board!). If this happens, follow the steps found in Appendix A.

Now that the Circuit Playground Express board is ready, it's time to code in Python. First, though, you need to make a decision. The process of creating code files involves typing in text, much like the JavaScript we saw in MakeCode. What software should you use to create the code files? Python commands (code) will be saved into a text-based file. For that we need a text-editing program.

TEXT EDITORS

Unless you are typing in the REPL (more on using the REPL later in the chapter), you'll want to type your CircuitPython code in a *text editor*. A text editor is a program on a PC, Mac, Linux, or Chromebook that accepts text input and allows you to edit the text and save the final result to disk.

No one wants to type complex programs into the REPL over and over. One typing mistake and we'd have to start all over. If we store the program text in a file, we can save it for later use.

Text editors come standard with all operating systems:

* PC/Windows: Notepad, Wordpad, Microsoft Word

* Linux: Nano, Vim, EMACS

* Mac: TextEdit, Pages, TextMate

* Chromebook: Caret, Google Docs, Writebox, Text

If you are in a learning environment, your teacher will often tell you which editor to use and guide you on how to use it. If you're going through on your own, you are free to select your own text editor. Full-fledged word processors such as Microsoft Word, Google Docs, and Mac Pages do not save plain text by default. A more basic editing program often works best.

Try the Mu Editor (if possible). Mu is a simple editor that runs on Windows, macOS, Raspberry Pi, and Linux (the list may expand to other platforms as the developers have time). A serial console is built right into the Mu program so that you get immediate feedback from your board's serial output and easily use the REPL in the same program!

If you find you cannot use Mu, use the text editor of your choice.

For Chromebook, the examples will be using the Caret editor. A separate terminal emulation program is needed to type commands into the REPL and to receive REPL and program output from CircuitPython. This book will use the Chromebook terminal emulator Beagle Term to perform serial input and output. Both applications are free in the Chrome OS app store.

> NOTE Mu will be shown in most examples. Mu is the recommended editor for Windows, Mac, and Linux. Please consider using it (unless you have a favorite editor already!). This eliminates the need for using two programs, a text editor and a terminal emulation program, to interact with the Circuit Playground Express serial input and output. Mu is not required. If you are more experienced, another editor and a terminal program will work fine.

> WARNING Ensure that you use an editor that writes out files completely when you save a file to disk. It is so easy to write your code in an editor and fail to save the code to disk. Many "good" editors will prompt you to save your work, but some do not. Both Notepad (the default Windows editor) and Notepad++ can be slow to write. You need to click the save icon to ensure your data is saved, and you must be sure to eject the drive.

EJECT OR SYNC THE DRIVE AFTER WRITING

If you are using a problematic text-editing program, not all is lost! You can still make it work.

On Windows, you can eject or safely remove the CIRCUITPY drive by right-clicking the drive in File Explorer and clicking Eject. It won't actually eject, but it will force the operating system to save your file to disk. On Linux, use the `sync` command in a terminal to force the write to disk.

When a program edit is complete, save a copy somewhere safe, such as a hard drive or data flash drive, to ensure that you have a copy for later use.

Installing Mu on Windows or Mac

The first step is to download the latest version of Mu. If you are using Windows, you must be running Windows 7 or higher. For macOS you must be running 10.12 (Sierra) or higher (Mac users with lower versions can try the Linux instructions that follow, but that is not guaranteed to work, according to the author of Mu).

The main Mu repository is located on the web at *http://codewith.mu*.

Select the latest version for your operating system. Download and save the Mu installation file to your desktop, download folder, or wherever is handy. Run the installation program. The installation process is operating system dependent (Figure 6-6). Windows installation programs ending in `.exe` or `.msi` can be run by double-clicking. macOS has its own install package.

FIGURE 6-6. The Mu program icon on Windows (left) and Mac (right) when placed on the desktop. Icons are subject to change by the Mu developers.

Once you have the program installed on your computer, you're ready to start coding Python.

Installing Mu on Linux

Each Linux distro is a little different, so use the following as a guideline. See *https://codewith.mu/en/howto/install_with_python* for details.

1. Open a terminal window.

2. Mu requires Python version 3. If you haven't installed Python yet, do so via your command line using something like `sudo apt-get install python3`.

3. You'll also need pip3 (or pip if you have only Python 3 installed); try running `pip3 --version`. If your system does not have pip installed, run `sudo apt-get install python3-pip`.

4. Finally, run `pip3 install mu_editor`.
   You can now run Mu directly from the command line.

USING MU

Mu attempts to automatically detect a Circuit Playground Express plugged into a computer. Before starting Mu, please plug in your CircuitPlayground Express and make sure it shows up as a CIRCUITPY drive in your computer's file explorer.

Once Mu is started, you will be prompted to select your mode (Figure 6-7). Please select Adafruit CircuitPython.

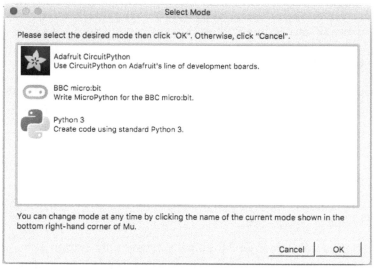

FIGURE 6-7. The Mu Select Mode screen

If you get a warning that the Circuit Playground Express can-not be found (Figure 6-8), ensure you have the board plugged into your computer. Check to see if the CIRCUITPY drive shows up in the available disk drives on your computer. If you are still having issues, see Appendix A for troubleshooting tips.

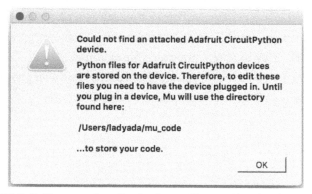

FIGURE 6-8. The warning that's shown if your Circuit Playground Express is not plugged in when Mu is started

You should now see the main Mu screen (Figure 6-9).

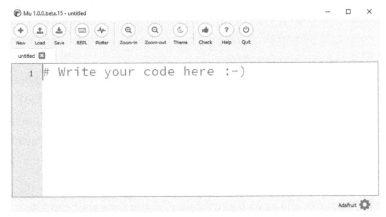

FIGURE 6-9. The Mu main editing window

Now that Mu is available, it is time to start coding Python.

CREATING PYTHON CODE

Plug your Circuit Playground Express via a known good USB cable into your computer. Your file explorer should show that a new flash drive is available named CIRCUITPY. (If you do not see the drive, see the earlier section "Downloading the Latest Version of CircuitPython" or Appendix A.)

Start the Mu editor. Usually you double-click the Mu icon in Windows or click the Mac icon. If you are on a Chromebook, run Caret or the editor you have chosen.

As with MakeCode, a good first program to create is one that makes a light on the board blink. Here are the general steps:

1. You set up the board environment before starting the main `forever` loop.

2. You create a loop that continually runs—in this case, turning an LED on, waiting, turning the LED off, waiting, then rerunning the loop.

Generic Python code does not know it will be running on a Circuit Playground Express. To assist Python with hardware-specific tasks, the creators of Circuit Playground Express have easy-to-use code that performs operations on the board, such as performing digital input and output.

In the larger Python world, thousands of modules and libraries are available for performing a wide variety of useful tasks. For the smaller CircuitPython, memory size limits the number of libraries a bit. Adafruit is committed to providing a wide variety of functionality for their CircuitPython products.

Why use a library? Libraries are great for two reasons:

* Libraries allows code portability. If you take your CircuitPython code for Circuit Playground Express and place it on another CircuitPython product such as an Adafruit Feather M0, the code will run (if you haven't used board-specific capabilities). The underlying hardware may be very different, but the authors of the library have done the translation from Python to hardware coding.

* Libraries allow a person to focus on their project and not the nuts and bolts of a specific hardware architecture. It is but one premise of open source software: someone has taken the time to write (hopefully useful) code to do something you would also like to do, such as code that makes a motor turn on and off. Coding the low-level motor control via hardware can be rather difficult. If the project designer can import a library to easily code something like `motor.on` and `motor.off`, the designer can focus more on a project and not have to intimately know how to control the motors on the circuit level.

The programmers at Adafruit have written a library for CircuitPython to interact with the Circuit Playground Express board's built-in hardware. It is called `adafruit_circuitplayground`.`express`. The library has a number of useful functions, which we will use in examples in this chapter.

Installing the Adafruit CircuitPython Libraries

Adafruit and contributors have created a large number of libraries to handle the built-in hardware on Circuit Playground Express and other devices that may be added onto the board (like a display, for example). You can find the file necessary to add this capability here: *https://github.com/adafruit/ Adafruit_CircuitPython_Bundle/releases*.

You will need to determine the CircuitPython release you are running on your board, which you can do by opening the `boot_out.txt` file on the CIRCUITPY flash drive when you plug the board in. As of this writing, there were versions for a 3.0.0 branch and a 3.0.0 branch. Since I am running the 3.0.0 version of CircuitPython, I chose the corresponding 3.0.0 zip file and saved it on my computer.

Open the zip file and you will see many small files used for a variety of input and output devices.

With the zip file open on your computer, highlight and copy the **lib** directory (which contains all the files and subfolders we need) to the CIRCUITPY drive. It will take some time and consume about 400KB of space. This still leaves plenty of room on the drive for programs. Be sure to use the **eject** function of your file explorer program (**sync** in Linux) to ensure all writes are made to the Circuit Playground Express flash memory before disconnecting the board from USB.

Now Circuit Playground Express should be set to use a wide variety of prebuilt libraries. If you ever need to refresh the library with the latest code, follow the same procedure.

To use the entire code library, you include the following command at the top of a Python program:

```
import adafruit_circuitplayground.express
```

Whenever you want to use a function in the module, you will need to type the function name, such as when reading the A Button:

```
button_read = adafruit_circuitplayground.express.button_a
```

Fortunately, we can shorten the function calls due to some Python, behind the scenes. Here is how to refer to the `adafruit_circuitplayground.express` module by using the handy acronym `cpx`:

```
from adafruit_circuitplayground.express import cpx
# Objects to Circuit Playground Express objects can now be
# referred to their abbreviated form
button_read = cpx.button_a
```

Much better to read—and less typing.

For the blinking D13 LED, we'll use the Circuit Playground Express library in CircuitPython.

Type the following code into your text editor:

```
import time
from adafruit_circuitplayground.express import cpx
while True:
    cpx.red_led = True
    time.sleep(1)
    cpx.red_led = False
    time.sleep(1)
```

For the lines after `while True:`, you should *indent* the code. Indenting (putting space before the text on a line) is done by either typing four spaces or pressing the Tab key. This informs Python that the code should all be contained in the statement above it. The `while True:` provides a `forever`-style loop as in Make-Code or the `loop()` function in Arduino. The value `True` will always be true so `while True:` will always loop all the statements within it.

To Tab or to Space

Some coders prefer to indent code using spaces. Many Python programmers indent with the Tab key, which is equal to a set number of spaces. Neither is wrong, but all agree that you should not mix tabs and spaces or you may end up with a Python error that will not be obvious to fix. Some text editing programs might put an actual tab character into the text file. The editor might convert a tab to multiple spaces. Consistency is the important thing. Using Tab will almost never make someone later editing your code question your choice.

Now save your program to a disk. You will want a place to store a copy of your code for later use; this can be a hard disk, a flash drive, or online storage like Google Drive.

When you save the program, use the filename blink.py. The .py at the end is called the file extension, and it lets you and others know the text file contains Python code. Also, save the program as code.py. This is the name CircuitPython recognizes as the current program you wish to run.

Program Names

The CircuitPython file you copy over to run on Circuit Playground Express should always be called code.py. This is so the Python interpreter knows the name of the code you want to run.

If you use mycoolprog.py or anything else, the board will not know that file is the code you wish to run. Feel free to use a more descriptive name on your backup storage device copies—for example, music-on-tilt.py.

The CircuitPython authors state there are four filenaming options for the code the board will run: code.txt, code.py, main.txt, and main.py. CircuitPython looks for those files, in that order, and then runs the first one it finds. The author and Adafruit suggest using code.py as your code file.

Still, it is important to know that the other options exist. If your program doesn't seem to be updating as you work, make sure you haven't created another code file that's being read instead of the one you're working on.

Throughout the book, we will use code.py as the CircuitPython program to run.

Running the Python Code

Okay, so at this point you have your Python program. Let's get that code running.

Plug in your Circuit Playground Express, via USB, to your computer. The board should show up as a flash memory drive named CIRCUITPY.

Copy the file `code.py` you saved earlier via your computer's file program over to your CIRCUITPY drive.

If you are using the Mu editor, you can use the save button to save the file to Circuit Playground Express if you named the program `code.py`. If you need a "Save as" function to make the copy, double-click the filename on the Mu tab that contains the program name. A dialog box will appear allowing you to name and save the code wherever you wish.

The red D13 LED next to the USB connector should now be flashing. The Python code is working!

On the off chance your code is not working:

* Be sure the green power LED is on and you see the CIR-CUITPY flash drive.

* Be sure your file is called `code.py` (and not `blink.py` or something else). Unlike MakeCode filenames, the `py` file on the board must be named `code.py`.

* You can press the Reset button and that should restart the Python code if it is not running already.

* Double-check your Python program against the earlier listing. The indented text should not mix tabs and spaces. You must indent the text as shown. In Python, that indicates the text is in a loop like MakeCode.

EXAMINING THE CIRCUITPYTHON BLINK CODE

Without going into the details of the entire Python programming language, we can understand what the code is doing by comparing the code to similar actions in MakeCode.

Modules are similar to the code blocks in MakeCode that allow you to select blocks of a specific type—for example, LOOPS for blocks that perform looping actions, LIGHT for blocks that turn LEDs on and off, and so on. What we do not have at hand are the range of statements that can be created in each library—the library building blocks. Let's explore them.

The `while True:` statement is the same as the `forever` loop in MakeCode. The code within the `while True:` loop will be executed "forever." The code that follows `while` can be any mathematic statement that evaluates to a `True` or `False` condition. It could be `x < 3` or another comparison that is evaluated as `True` or `False`. Here we just use `True`, so in essence `while True:` always loops (it never exits because the constant value `True` is never `False`).

The code in the `while` loop is similar to the code used in MakeCode. The LED is turned on by setting the value of the object `cpx.red_led` to `True`. The time function waits 1 second. The LED is turned off by setting the value of the object `cpx.red_led` to `False`. Then another second elapses before starting the loop again.

With additional examples, you'll see more Python statements and library functions that build on the work of the blink example.

Using the Internal (Frozen) CPX Library

The CPX library is always available to CircuitPython programs. As of CircuitPython version 2.3.0, CPX is a *frozen library*. A frozen library is part of the core of CircuitPython as of version 3.0.0 and higher. No file in the Circuit Playground Express /lib directory is required to support the CPX library.

If you have upgraded from a CircuitPython version prior to 3.0.0 and you have library files in the /lib subfolder on the CIRCUITPY drive, you will want to delete the CPX library in /lib. You do this by deleting the directory and files in /lib/adafruit_circuitplayground/. That way, you can be sure that you are using the frozen version of the CPX library.

OUTPUT FROM CIRCUIT PLAYGROUND EXPRESS TO THE COMPUTER

Earlier when we programmed Circuit Playground Express using MakeCode, there were no blocks that allowed the user to interactively communicate with the board. The computer allowed you to code and let you see what that code would do. But there was no ability for the code to take computer input or format output to the computer. Fortunately, CircuitPython has greater flexibility when performing input and output.

Up until now, the USB connection has provided two functions:

* Power the board via the USB cable

* Enable loading code and reviewing files on the device as if the board is a flash memory drive

Universal Serial Bus (USB) provides a number of other functions that are very useful. This capability can be used by programming in CircuitPython (or later in Arduino). Two additional functions that are extremely useful are

* Input and output to the connected computer over USB—typically called serial communications

* Human Interface Device (HID) mode, which allows the board to emulate devices such as a keyboard or a mouse

Serial Output

Time to modify our first Python program. Add the line that starts with `print` below the `while True:` statement to your code:

```
import time
from adafruit_circuitplayground.express import cpx
while True:
    print("Hello CircuitPython!")
    cpx.red_led = True
    time.sleep(1)
    cpx.red_led = False
    time.sleep(1)
```

Please make sure it is indented and typed correctly with two double quotation marks and two parentheses.

The `print` statement will send the function argument to the serial output (out the USB port back to your computer).

Here is where using the Mu editor helps. If you do not have Mu, skip ahead for how to get the serial output.

Click the Serial icon, the fifth icon from the left with a double arrow icon on it (see Figure 6-10).

FIGURE 6-10. The Serial icon in the Mu editor, fifth from the left, has a double arrow icon.

The Serial screen acts both as a command input/output window and as an interactive REPL Python environment.

Once you have typed in the print line, your screen should look similar to the one in Figure 6-11. You can adjust the size of each window; in Figure 6-11, I expanded the Serial window to show more of what is being output by Circuit Playground Express. Left-click and hold the mouse on the light gray line between the code and Serial windows, adjust, and then release the left mouse button.

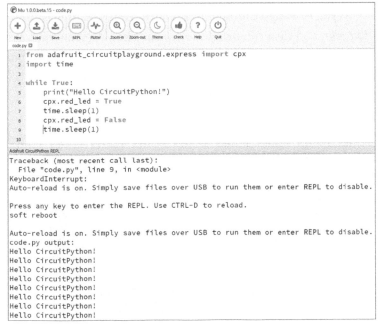

FIGURE 6-11. With the Serial window open, serial output can be seen.

Every two seconds, you should see the words *Hello, Circuit-Python!* display on the screen. If you don't see the text at the bottom of the screen, press Ctrl+D to restart the program.

The print function allows the user to print text and numbers through the USB port on Circuit Playground Express to the host computer. With the Mu editor, the text shows up in the Serial window.

If you are not using Mu, the text can be seen by any program that opens Circuit Playground Express as a serial device. On a Mac, the screen command opens up a compatible serial connection. On Windows, you need a terminal emulation program such as PuTTY. On Chromebooks, you need a terminal program such as Beagle Term.

If you're using a Chromebook, plug in your Circuit Playground Express, and then run Beagle Term. You should get a screen similar to the one in Figure 6-12 with a /dev/ttyACM0 or ACM1 port already filled in. If you need to switch ports to a different USB device, do it here. Then click the Connect button (the other settings are fine).

FIGURE 6-12. Beagle Term on Chromebook, settings screen

Figure 6-13 shows the corresponding Beagle Term output from the CircuitPython print statement.

Once you have a method for getting data from Circuit Playground Express to your computer, what data might be output?

FIGURE 6-13. Chromebook serial output to Beagle Term for the Hello CircuitPython demo

With all the sensors on the board, it would be great to use Circuit Playground Express to measure things like temperature and light intensity and send them back to the computer to record for later use. The larger NeoPixel LED lights can indicate certain values measured by the board, but actual numbers for values makes the output easier to use for measurements. You can do so easily (and you will in the next chapter) when you learn more about functions available for program use.

The Circuit Playground Express code library has the functions that let you use all the capabilities of the board. The next section will list them all for reference.

THE ADAFRUIT CIRCUIT PLAYGROUND EXPRESS LIBRARY

To effectively use the capabilities of Circuit Playground Express in CircuitPython, we need a reference for all the library functions available. This section lists the functions available as of this writing. An up-to-date list (for example, if Adafruit changes or adds to the library) is located at *http://circuitpython.readthedocs.io/ projects/circuitplayground/en/latest/api.html*.

All the code fragments in Table 6-1 assume you have the Python `import` statement at the top of your code like this:

```
from adafruit_circuitplayground.express import cpx
```

> **NOTE** A *parameter* is a value that you provide to a function to set the behavior of that function. For example, in the two statements `print(3.0)` and `print(x)` the value `3.0.0` and the variable `x` are considered the parameters used for the `print` function.

TABLE 6-1. Adafruit Circuit Playground Express library functions and usage

ACTION	CPX LIBRARY USAGE
Accelerometer	`x, y, z = cpx.acceleration` `print((x, y, z))`
Button A	`if cpx.button_a:` `    print("Button A pressed!")`
Button B	`if cpx.button_b:` `    print("Button B pressed!")`
Slide Switch	`print("Slide switch:", cpx.switch)` (True to the left, False to the right)

Red (D13) LED	`cpx.red_led = True`
	`time.sleep(1)`
	`cpx.red_led = False`
	`time.sleep(1)`
	(True turns LED on, False turns LED off)

Temperature Sensor Value	`temperature_c = cpx.temperature`
	`temperature_f = temperature_c * 1.8 + 32`
	`print("Temperature celsius:", temperature_c)`
	`print("Temperature fahrenheit:", temperature_f)`

ACTION	CPX LIBRARY USAGE
Light Sensor Value	`print("Light Value:", cpx.light)`
Play WAV file	`cpx.play_file("laugh.wav")` The WAV file must be on the board flash drive.
Play a tone on the speaker (fixed duration)	Parameters: `frequency` (*integer*)—The frequency of the tone in Hz `duration` (*decimal*)—The duration of the tone in seconds `cpx.play_tone(440, 1.0) # 440 hz for 1 second`
Play a tone on the speaker (until told to stop)	Parameter: `frequency` (*integer*)—The frequency of the tone in Hz `cpx.start_tone(262)`
Stop playing a tone previously started with `start_tone`	`cpx.stop_tone()`

Detect the board being tapped	`cpx.detect_taps = 1`
	`if cpx.tapped:`
	`    print("Single tap detected!")`

Detect the board being double-tapped	`cpx.detect_taps = 2`
	`if cpx.tapped:`
	`    print("A double-tap detected!")`

Detect when board is shaken	Parameter: `shake_threshold` (*integer*)—The threshold shake must exceed to return `True` (default: 30). Lower = more sensitive; keep above 10.
	`if cpx.shake():`
	`    print("Shake detected!")`
	`if cpx.shake(100):`
	`    print("Hard shake detected!")`

Set NeoPixel LEDs	Set the color of NeoPixels; values are Red, Green, Blue, and can each range from 0 (off) to 255 (full on): `cpx.pixels[9] = (30, 0, 0)`

ACTION	CPX LIBRARY USAGE
Set NeoPixel LEDs *(continued)*	Pixels are numbered counterclockwise from `cpx.pixels[0]` through `cpx.pixels[9]`.

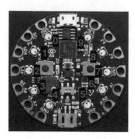

You can also use a hexadecimal value in the format `0xRRGGBB` (decimal 30 = hex `0x1e`):

```
cpx.pixels[9] = 0x1e0000
```

All the pixels can be lit to the same color value specified using `cpx.pixels.fill`:

```
cpx.pixels.fill( (30, 0, 0) )
```

To turn off all pixels, set them to `(0, 0, 0)`:

```
cpx.pixels.fill( (0, 0, 0) )
```

ACTION	CPX LIBRARY USAGE
Set NeoPixel brightness	Set the brightness of all pixels (from `0.0` to `1.0`): `cpx.pixels.brightness = 0.3`
Pads A1 through A7 being touched	The touch pads are around the edge of the board.

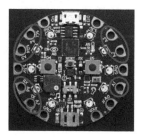

```
if cpx.touch_A1:
    print('Touched pad A1')
```

Change A1 to A2, A3, A4, A5, A6, A7 for other pads.

ACTION	CPX LIBRARY USAGE
Set touch pad sensitivity	Parameter: adjustment (*integer*)—The desired threshold increase; higher numbers make the touch pads less sensitive.

```
cpx.adjust_touch_threshold(200)
while True:
    if cpx.touch_A1:
        print('Touched pad A1 hard')
```

CircuitPython API Documentation

You can get the information for using the libraries built to implement Circuit Playground Express–specific hardware control functions at *http://circuitpython .readthedocs.io/.*

The code that makes up how our program communicates with a predefined set of code written by another group is often called an applications programming interface (API). If you like, you can look at the source code for Circuit Playground Express library functions in the Adafruit GitHub repository at *https://github.com/ adafruit/Adafruit_CircuitPython_CircuitPlayground/ blob/master/adafruit_circuitplayground/express.py.*

Go ahead and try some of these functions in your own code. To get ideas on how others are using CircuitPython on Circuit Playground Express, visit *https://learn.adafruit.com/category/express* and look for projects that are using CircuitPython.

RUNNING CODE ON EXPRESS VIA THE REPL

Earlier in the chapter, we used the Serial window in Mu (or Beagle Term for Chromebook) to view the output from a CircuitPython file. Within the Serial window, you can press Ctrl-C to get to the REPL. The REPL is an interactive method for entering commands into CircuitPython and getting feedback.

First, connect your Circuit Playground Express board to your computer with a USB cable. Run the Mu editor (Windows/Mac) or Beagle Term (Chromebook).

If all is good, you will see the editor window shown in Figure 6-14. Click the REPL button, which has a keyboard icon.

FIGURE 6-14. Start the Serial by clicking the Serial icon at the top of the Mu screen and press Ctrl-C to get the >>> prompt.

The editor window will split in half (Figure 6-15). The REPL is in the bottom portion.

The Serial window will show your serial output/input. But it will also communicate with the board. If you press Ctrl-D, the code will start again without doing a full board reset (which pressing the button onboard does).

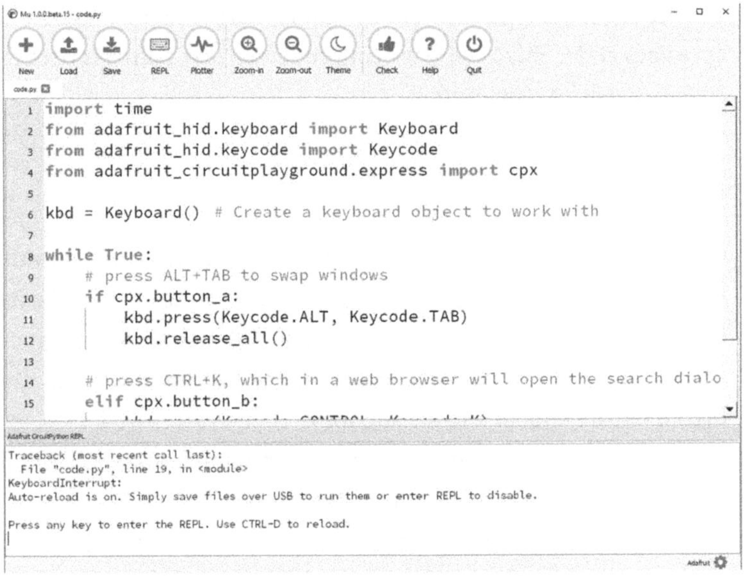

FIGURE 6-15. The Serial window at the bottom of Mu

If you press any other keyboard key, you enter the REPL itself (Figure 6-16). Now any commands you type into the window will be interpreted as CircuitPython commands.

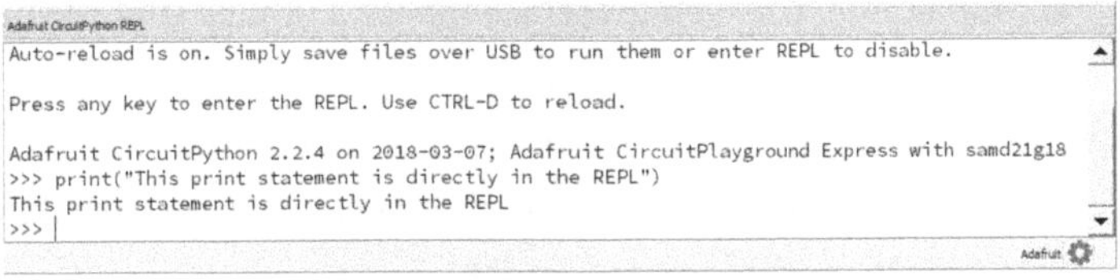

FIGURE 6-16. Typing CircuitPython commands into the REPL

If you would like to import a library (Figure 6-17), you can do so first and then you can use library functions after that.

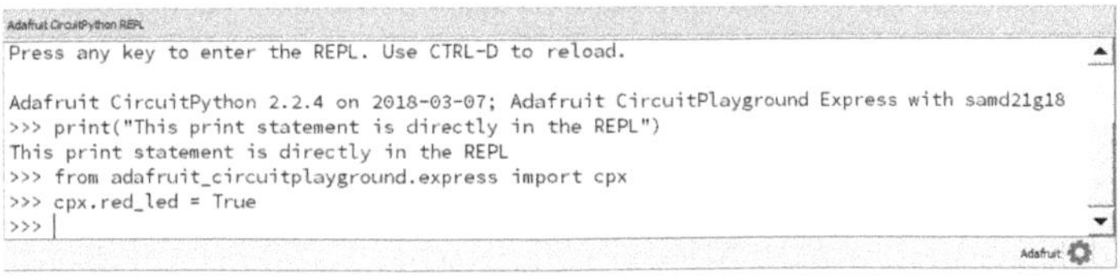

FIGURE 6-17. Importing a library in the REPL

The code you type is interactive, but there is no mechanism to save your code. This is why writing code in a text editor is best for code you will be changing.

WRAP-UP

CircuitPython provides a great way to program Circuit Playground Express using a programming language gaining in popularity. CircuitPython also exposes additional functionality on Circuit Playground Express, including the ability to read and write files placed on the onboard flash memory.

You can also type CircuitPython commands into the REPL if you need to perform a short list of actions.

In the next chapter, some of the more advanced uses of CircuitPython for Circuit Playground Express will be covered.

CHAPTER QUESTIONS

1. What is an interpreted computer language?

2. In MakeCode, the web interface places a binary UF2 file onto Circuit Playground Express when the download button is used. How is code placed onto Circuit Playground Express in CircuitPython, and where is the mechanism to turn code into binary machine commands?

3. What is the CircuitPython equivalent command to the MakeCode `forever` loop?

4. Which CircuitPython function outputs information from Circuit Playground Express to the connected computer?

7

Using the Circuit Playground Express CircuitPython Library

Writing Circuit Playground Express programs that interact with the board's features is greatly simplified by using the specialized CircuitPython library provided by Adafruit. The library, introduced in Chapter 6, allows project creators to focus on their goals and not spend unnecessary time figuring out the internals of the board.

In this chapter, the Circuit Playground Express CircuitPython library will be used to read the sensors on the board. The data is first displayed in the Serial window and then to a spreadsheet. Finally, the board's exterior pads are set up for capacitive touch switching, letting you coax music out of fruit from your refrigerator.

READING SENSORS

In Microsoft MakeCode, reading sensor values involved pulling in the correct blocks. But MakeCode does not really provide for an easy way to manipulate data over time or to put the data somewhere for later use.

CircuitPython excels at data manipulation. From simple data gathering to complex calculations and plotting, CircuitPython is the language to choose.

To demonstrate the abilities of CircuitPython, the following program will get things started. The program will

1. Read the temperature in Celsius.

2. Read the light intensity from the light sensor.

3. Output the data to the serial output connection to the host computer.

4. Wait a bit.

5. Do everything over again.

Looking at the list of library functions in the `adafruit_circuitplayground.express` module in the previous chapter, the value used to read the temperature in Celsius is `cpx.temperature` and the value used to read the light intensity is `cpx.light`. Thank you to the library authors for handling all the low-level interfacing to the hardware!

Libraries Installed?

If you did not install the Adafruit libraries to the Circuit Playground Express flash memory subdirectory /lib, discussed in Chapter 6, please go back and do so now. The libraries will be used extensively in this chapter.

First import the modules. You need the `adafruit_circuitplayground.express` module to read the sensors, and you need the `time` module to use the `sleep` function.

```
import time
from adafruit_circuitplayground.express import cpx
```

Next we need to set up a `forever` type loop that takes the light and temperature readings, prints them out to the serial output, waits a set period of time, then loops back to do it again.

```
import time
from adafruit_circuitplayground.express import cpx
while True:
    print("Temperature:", cpx.temperature, ", Light
Intensity:", cpx.light)
    time.sleep(1)
```

Type this code into your text editor, such as Mu or Caret. Save a copy on your computer as **temp-light.py**. Then save the code on your Circuit Playground Express CIRCUITPY drive as **code.py**.

In the Mu REPL window, you should see the program's serial output in Figure 7-1. On a Chromebook, similar output should be visible in Beagle Term.

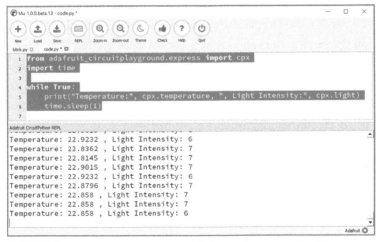

FIGURE 7-1. The temperature and light code running in Mu

The temperature is in degrees Celsius and the light intensity is a relative level. Note the light sensor has not been calibrated on Circuit Playground Express; you should look for changes in light rather than think of the values as accurate readings.

Go ahead and place a finger over the temperature sensor on the board (see Figure 7-2).

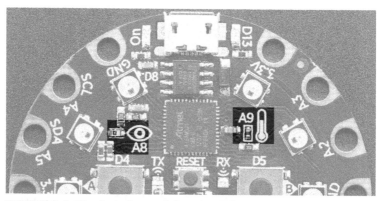

FIGURE 7-2. The locations of the Circuit Playground Express light and temperature sensors

The values for temperature should go up due to body heat being usually warmer than the room temperature. If you release the sensor, the temperature should gradually go down (if you blow on the sensor, the values may go down more rapidly).

Now use your hand to shade the light sensor. Covered up, it should register 5 or below. In fluorescent light, the reading should be about 8 or 9. Holding the sensor to a white portion of the computer display should provide a reading over 20.

Reading the Circuit Playground Express sensors and writing the output is perfect for scientific sensing and analysis.

Your Turn: Sensor Values

Change the code in the following ways:

* Instead of temperature and light, print out the three values for the accelerometer chip in the center of the Circuit Playground Express. The chip shows acceleration (changes in movement speed). With a `print` statement displaying the `cpx.acceleration.x`, `cpx.acceleration.y`, and `cpx.acceleration.z` components, move and shake the board to watch how the values change. Describe how these x-, y-, and z-axis values vary by what type of movement you make. Look at Circuit Playground Express in the lower-left corner of the central accelerometer chip to see a diagram describing which motion changes which value. If you would like to try this interactively in Mu, skip to the end of the next section.

* The accelerometer action is simplified by the `shake` function. Print out the `True` or `False` returned by `cpx.shake`. After you are familiar with what the shaking of the board will produce as far as `True` or `False`, adjust the sensitivity of the shake by including a number parameter, for example, `cpx.shake(50)`. Adjust the number to see how it affects the shake detection.

* Diving deep: Interested in sound and how you might code a CircuitPython implementation of a VU sound meter? See the tutorial at *https://learn.adafruit.com/adafruit-circuit-playground-express/playground-sound-meter* to learn more.

FILE INPUT AND OUTPUT

Printing Circuit Playground Express sensor data to the serial output (viewable via the Mu Serial window or a serial terminal like Beagle Term) works well for recording data, but this process is far from convenient. The data may be captured via a screen shot, but a picture is not that helpful. The data may also be available via the operating system "cut-and-paste text" capability. But it probably will not be formatted well and is inconvenient to use long term.

The easy, professional way to capture data from Circuit Playground Express is to have the code write output to a file that can be used on the computer later. Writing out to a file also means you do not have to physically be there to watch the results of an experiment to record results. A CircuitPython program can take a Circuit Playground Express sensor measurement every 10 minutes for an hour. After all, we have a programmable microcontroller—have it do the tedious work.

The question for the past 70 years in computing is: How should the data be written to a file to have it easily read back again? Text files with easily read text would be the best format for information to be used in word processing. But text is not well suited to processing numeric data, a problem solved 35 years ago with the creation of VisiCalc, the first spreadsheet. Today, spreadsheets are used for both tiny data sets and vast amounts of data. Your computer likely can use Microsoft Excel, Google Sheets, Apple Numbers, or an open source suite such as Apache OpenOffice or LibreOffice.

Most spreadsheets read and write data in their own, *native* (proprietary) binary format. Even if we knew these exotic data formats, coding CircuitPython to read and write the proprietary formats would be laborious and redundant.

Fortunately, all spreadsheets do understand the *comma-separated values (CSV)* file format, a text-based format from the early 1980s. CSV output can be read by many programs, including spreadsheets, data presentation programs, databases, and word processors.

The typical format of a CSV file consists of values separated by commas with a return/newline/enter character at the end of each line. Take the temperature and light intensity readings from the previous section. This is how the data might look in CSV format:

```
Temperature, Light Intensity
22.9232, 6
22.8362, 7
22.8145, 7
22.9015, 7
22.9232, 6
22.8796, 7
22.8580, 7
```

Each set of values is separated by a comma. The first line is the titles of the columns of numbers. (Sometimes double quotes are put around each value, but that is not usually necessary.) Each set of values is on its own line of text. This is just one way to write out the data, but it is a format that most programs easily digest.

Currently, when CircuitPython boots up, it sets the files on the built-in flash drive such that the operating system can read and write files and programs can only read files but not write a file. *Read-only* is the term used when a program can read files but not write them. If you were to try to write a file in a Circuit-Python program when the CIRCUITPY drive was in the read-only file mode, your program would stop with an error.

You can set the CIRCUITPY flash storage to writable by a program. Just create a short program named boot.py and put it in on the drive (in the main or root directory):

```
# boot.py
# Set Circuit Playground Express flash chip to program
# writeable
#       If toggle switch is right,
#       flash is program writeable and file access is frozen
#       If toggle switch is left,
#       flash chip file access ok, file writes give an error
#       https://learn.adafruit.com/cpu-temperature-logging-
#       with-circuit-python/writing-to-the-filesystem
#       writing-to-the-filesystem
```

```
# 2018 Mike Barela for Getting Started with Circuit
# Playground Express 2018 MIT License
import storage
from adafruit_circuitplayground.express import cpx
storage.remount("/", cpx.switch)
```

Save this code on your hard drive/permanent storage and to your CIRCUITPY drive as **boot.py**. Use your operating system to eject the drive (or pressing Reset once will do), then unplug your Circuit Playground Express from the USB cable and then plug it back in. The code allows a program to write to the flash drive (write file access) if the slide switch is to the right (with the USB port up). If you wish to remove the file write capability of the flash drive, move the slide switch back to the left, press the Reset button, and see Figure 7.3 for the slide switch settings.

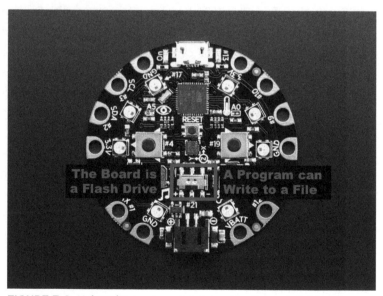

FIGURE 7-3. Using the boot.py program to allow switching the flash drive from filesystem access to program read/write access

Switching the Filesystem Write Capability

Unfortunately, changing the flash to allow a program to write a file sets the flash drive so you cannot change files using your computer operating system! It is either-or. This is inconvenient but manageable using some boot.py code to use the Circuit Playground Express onboard toggle switch. The following graphic shows the error in Windows when you try to copy a file to the CIRCUITPY drive when the drive has been set to read-only for file access via boot.py.

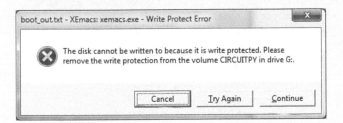

boot.py runs only on first boot of the device. If you change the toggle switch setting, you'll have to press the Reset button (or eject the **CIRCUITPY** drive, then unplug Circuit Playground Express from power and plug it back in).

If you want to use the toggle switch for an application that interferes with its use for switching the filesystem write status, remove boot.py or rename it to something else. If you need both the toggle switch and filesystem switching, another switch method should be considered such as detecting whether push button A or B or both are pressed on boot.

The modified CircuitPython program that provides for CSV data output follows. Note that it is based on the temperature and light example we used earlier, but it could just as well be any measurement you want to make with your Circuit Playground Express with the proper changes.

```python
# Read Temperature and Light Intensity, output as a CSV file
# Mike Barela for Getting Started with Circuit Playground
# Express 2018 MIT License
import time
from adafruit_circuitplayground.express import cpx
# Set NeoPixel 0 to green (status), NeoPixel 1 to collecting
# data
cpx.pixels[0] = (0, 90, 0) # coded red, green, blue cpx.
pixels[1] = (0, 0, 90) # Pixel 1 blue when collecting data
num_readings = 10 # set to any finite value you want
# we try to open/create the file for append access and write
# the heading line. If an error occurs, go to except try:
try:
    with open("/temp-light.csv", "a") as fp:
        fp.write('Temperature, Light Intensity\n') # headings
        for x in range(0, num_readings):
            temp = cpx.temperature
            fp.write(str(temp) + "," + str(cpx.light) + "\n")
            # Change the value of sleep time below in seconds
            # 1 minute=60 sec, 5 mins=300 sec, 1 hour=3600...
            time.sleep(1)
            if cpx.button_a:
                break
        # Done, set NeoPixel 1 to green also
        cpx.pixels[1] = (0, 90, 0)
except OSError as e:
    # set NeoPixel 1 off and blink NeoPixel 0 (status)
    #       depending on the OS error
    cpx.pixels[1] = (0, 0, 0)    # Blank NeoPixel 1
    message_color = (99, 0, 0)  # Red for problem
    if e.args[0] == 28: # Device out of space
        message_color = (228, 160, 40) # set to Orange
    elif e.args[0] == 30:       # Device is read only
        message_color = (181, 90, 0) # set to Yellow
for x in range(1, 10): # Flash 10 times
    cpx.pixels[0] = message_color
    time.sleep(1)
    cpx.pixels[0] = (0, 0, 0)
    time.sleep(1)
```

Note the new use of the `try/except` blocks. Within the `try` block, we have placed code that might produce a Python error at runtime. With files and operating systems, a number of errors may crop up. The disk you are attempting to write to may be read-only. Or there may be so many files on the disk that there is no free space for more data. The `except` block contains code that will handle errors to inform the user there is a problem. In the previous code, the error type (in the strangely named variable `e.args[0]`) is tested to see if there are disk errors. If the errors are Out of Space or Read Only, the NeoPixel colors are set to Orange and Yellow, respectively, to let the user know the specific error. The value Red is used if any other error is detected. Similar error testing can be used for other errors, as noted in Python reference materials.

Edit the program to ensure you are collecting the number of data points you want at the time interval you want. Perhaps start with a short time interval, like one second, as a test; then later you can extend it.

Save the program on your permanent storage with a name like **temp-light-csv.py**. Then save a copy onto your Circuit Playground Express CIRCUITPY drive as **code.py**. Unplug your board and plug it back in again.

If your board shows a red #0 NeoPixel, then the program ran into a general problem. Orange indicates your flash drive is full. Yellow indicates you need to put the `boot.py` program onto your drive as noted earlier to enable write access and to ensure the toggle switch is set to the right, which is away from the speaker.

When the program is running, NeoPixel #0 should light up green, and NeoPixel #1 should be blue. The program is now collecting temperature and light values. When it is done, the second NeoPixel will turn green as well. If you set long time periods between readings or you collect a lot of data, the data collection could take quite a while.

When the program is done, you will have two green NeoPixels. Switch the slide switch back to the opposite value and press the Reset button once. You can now read and write files on the CIRCUITPY flash drive again.

Using your computer, you should see the file `temp-light.csv` in the main directory of the CIRCUITPY drive. You can view the file two ways:

* Open the file with your text editor. You will need to ensure you type the entire filename, `temp-light.csv`, and ensure you are pulling the file from the CIRCUITPY drive.

* If you have a spreadsheet program permanently loaded on your computer, selecting the `temp-light.csv` file (in Windows: double-left-click) will load the file into your spreadsheet program. If not, open your spreadsheet and specifically load the file on the CIRCUITPY drive. You might have to copy the file onto your computer drive (or the cloud for Google) if you have problems.

Figure 7-4 is a screenshot of Microsoft Excel when the `temp-light.csv` file is loaded. The data is in two spreadsheet columns, with row 1 containing the headings.

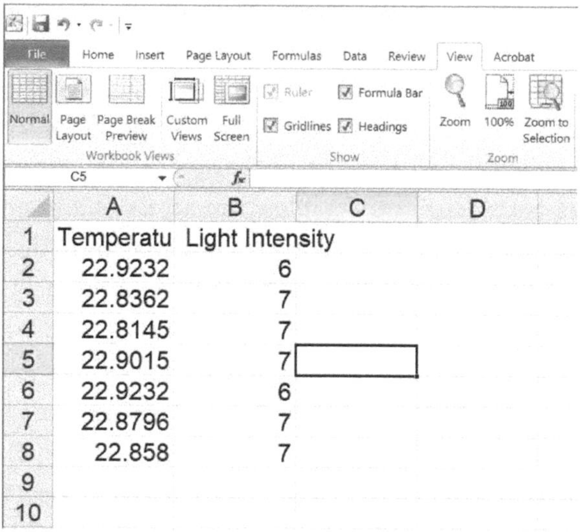

FIGURE 7-4. The `temp-light.csv` file loaded into Microsoft Excel

Loading the CSV file into other spreadsheets should provide a similar view. If you open the file in a text editor or word processor, you will see the values similar in appearance to a spreadsheet, but columns may not align and there will be commas between values.

Once you have the data in a file, you can easily use it. Spreadsheets usually have a single command to graph the output. Figure 7-5 shows the values graphed in Excel after I chose Insert → Line Charts → 2D Line Plot.

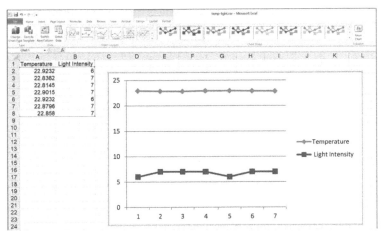

FIGURE 7-5. Graphing the temperature and light intensity values in Microsoft Excel

As you have seen, there's a bit more involved when you're working with the operating system using Python. Fortunately, Circuit Playground Express has encapsulated many of the "tough to do" work inside the library so that users can focus on their projects.

> NOTE When you decide to write other programs that do not require writing to the flash drive, consider renaming boot.py to something else, say boot-py.txt. This will "free up" the slide switch for other use.

Mu Plotting

Mu has an internal plotting capability somewhat similar to the data plotting done in the previous section. Why, you ask, was all the file work introduced? The Mu plot method is very handy for seeing what data is doing while things are actually happening but has the downside of not being able to save the data. But the plot mode is an excellent way to get instant feedback on what a sensor is doing.

First, one more Python data type is needed. A *tuple* is a group of several numbers or objects. A tuple can be used by Python as a whole unit or group. Usually, a tuple is enclosed in parentheses as follows:

```
()                      # An empty Tuple
(5, )                   # One item Tuple
(5, 3  )                # Two item Tuple
(5, 3, 8)               # Three item Tuple
(9, 2, 1, "spam")       # A four item Tuple
```

Mu uses data written to the serial output of a board to provide a real-time plot of those numbers on a graph compared to time.

Take the following sensor code, similar to previous examples; this code reads the three-axis accelerometer in the center of Circuit Playground Express. Three values are produced: the acceleration of motion in the x, y, and z directions. The values are written to the serial output as a tuple (within parentheses).

```python
import time
from adafruit_circuitplayground.express import cpx
while True:
    # Read motion sensor into values x, y, z
    x, y, z = cpx.acceleration
    # Print Tuple of acceleration values to serial out
    print((x, y, z))
    # Wait 1/10 second before getting a new value
    time.sleep(0.1)
```

Figure 7-6 shows a view of Mu with the Serial window open showing the serial output as a three-item tuple of decimal numbers. When we click the Plotter icon at the top portion of the Mu window, an additional window opens next to the Plotter and Mu automatically plots the tuple data. The plot in Figure 7-6 shows three values, which change over time as Circuit Playground Express is moved.

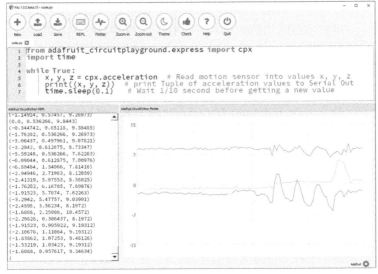

FIGURE 7-6. The motion sensor data output as a tuple with Mu plotting the data as the values are generated

If the data is not coming out to the Serial window, press Ctrl-D to reset the code.

You can modify the time between samples by changing the `time.sleep` value.

For getting a feel for what individual sensors can do, the Mu plot capability is wonderful. For saving data for later use, it is not as convenient as having data in a CSV file to plot yourself. Another factor is you have to use Mu, which is not available for Chromebooks.

Your Turn: Data Logging

To explore data logging:

* Choose which sensors you would like to monitor. Modify one of the programs we've looked at to output the sensor values you have chosen. If using the file method, be sure to select your own filename, ending in .csv. Does your data contain more than one or two values? Which lines of code are modified to output additional values? If using Mu plotting, be sure to print the parentheses and a comma if the tuple has less than two values, for example, (temp,) for a one-value tuple.

* How would the code change if you wanted to record a sensor value every half hour for a day? Note that Circuit Playground Express will "sleep" for long periods, seeming not to do anything; collecting data for science often requires patience to get a complete data set.

* Bonus: Add a column to record the time of each measurement. Hint: The time.clock() function returns the time since the board was last booted. Unfortunately, there is no real-time clock onboard to get an hour-of-the-day time.

* Extra Credit: boot.py uses the slide switch to inform Circuit Playground Express if the flash drive can be written to or if it is in "normal" operating system file access mode. Can you describe another way a user could inform boot.py which mode to set on boot? Note there are multiple ways to read an input on the board.

CAPACITIVE TOUCH AND MUSIC

Besides the three switches located on Circuit Playground Express, the outside pads marked A1 through A7 can switch states based on touch. The ability to touch an electrically conductive object and have it register as a change, which can be likened to a switch being activated, is called *capacitive touch*.

In electronics, capacitance is the ability to store electrical charge. It turns out human skin collects electrons fairly well. Walking across a carpet and then getting a shock when touching something is an example. You can rub a balloon on your hair and make the balloon stick to your head (or spike up your hair). This is because the balloon rubs some electrons off your hair, giving the hair a net positive charge while the balloon gathers a negative charge. Positive and negative charges attract, so the balloon and your hair stick together. In electronics, storing electrical charge is the job of a component called a *capacitor*. The amount of charge a capacitor can hold is called *capacitance*.

The Circuit Playground Express microcontroller has special circuitry onboard to read the change in capacitance of pads A1 through A7, and then register it. A library function can be used to read the pad like a push-button switch.

The next program uses the Circuit Playground Express library functions for capacitive touch and for playing tones from the onboard speaker. When you touch the A1 through A7 pads, the capacitance of your body is combined with the capacitance of the touchpad, thereby changing it. That change causes Circuit Playground Express to play different sounds. This code uses the Python if statement to test which pad was touched.

```
import time
from adafruit_circuitplayground.express import cpx
cpx.adjust_touch_threshold(50) # Set to be fairly sensitive
while True:
    if cpx.touch_A1:
        cpx.play_tone(440,1)   # Note A4
    if cpx.touch_A2:
        cpx.play_tone(494,1)   # Note B4
    if cpx.touch_A3:
        cpx.play_tone(523,1)   # Note C5
    if cpx.touch_A4:
```

```
        cpx.play_tone(587,1)   # Note D5
    if cpx.touch_A5:
        cpx.play_tone(659,1)   # Note E5
    if cpx.touch_A6:
        cpx.play_tone(698,1)   # Note F5
    if cpx.touch_A7:
        cpx.play_tone(784,1)   # Note G5
    time.sleep(1)
```

Try this program. When you touch the outside pads marked A1 through A7, it plays musical notes.

The `cpx.play_tone` function is used to play a one-second tone at a specified frequency. The frequency numbers used are not nice, round numbers, but they do produce the pure musical notes noted in the comments.

How do you know which frequency plays which pure musical notes? You look on the Internet. View *https://pages.mtu.edu/~suits/notefreqs.html* in your web browser to find a list of musical notes and their corresponding frequencies.

Feel free to select any musical note frequencies you like. Or you can select non-note frequencies, such as 100 through 700 hertz. In that case, you may notice the tones do not sound "pure," as they do with musical instruments.

You can also adjust the sensitivity of the touchpads up or down to your taste. The touch sensitivity may depend on the humidity in the air and other factors.

> NOTE Pads A1 through A7 can be used for capacitive touch. Why not pad A0? A0 is used for analog voltage output and so it cannot be set up for touch input.

It's simple to use the CircuitPython library functions to add touch sensing as something to check for, and then act on. But you are not limited to just touching the side tabs.

It turns out that the A1 to A7 pads can be extended using a *conductive material*. Conductive materials are usually metals, which allow for the free movement of electricity from one place to another. In electronics, this most often takes the form of *wires*—metal in long, cylindrical lengths. You can connect the bare end of wires to Circuit Playground Express; touching the ends should trigger the switch. It turns out that electricity likes a solid connection from one place to another, and wrapping a wire through the hole of the touchpad is not sturdy enough. For permanent connections, *solder* (a soft metal that melts with heat and cools to form a bond) is used. Given that Circuit Playground Express may be used for many projects, a less permanent connection method is probably best.

In comes *alligator clips*—tiny metal "jaws" with a spring inside to hold the two sides together (Figure 7-7). When squeezed, the sides open. They can be connected on the pad of Circuit Playground Express (the large hole in the pad helps make a good mechanical connection). See Figure 7-8. The other end of the wire has a sturdy pin.

FIGURE 7-7. Alligator clips with wires. Shown is Adafruit product number 3255.

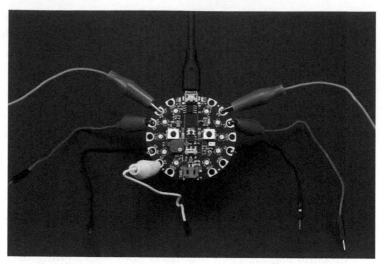

FIGURE 7-8. Alligator clip wires connected to capacitive touch-pads A1 to A7

> NOTE When Circuit Playground Express first powers on, the internal code *calibrates* the touchpads (sets them up for optimum performance). If you extend the pads with wires and other objects, you may need to press the Reset button for the code to recalibrate to the new project configuration.

At this point you can touch the end of the wires when you run the capacitive touch CircuitPython code and get the seven musical notes. But the wires still have not helped in making an easy-to-use project. For a musical instrument, like a piano, *keys* are ergonomically designed to fit human fingers to activate musical sounds.

You can use an index card or other stiff piece of paper to lay out piano-style keys (see Figure 7-9). Then you can use tape to secure the wire ends to the correct places, making sure the exposed wire can be touched by a finger. Much more ergonomic!

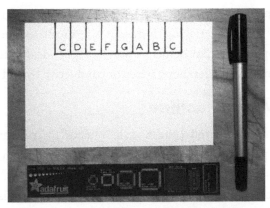

FIGURE 7-9. Laying out a piano-style keyboard for capacitive touch use

To make the contact area between the end of the wire and a finger, get a flat, conductive material to make a larger pad. Some folks have soldered copper pennies to the end of the wires. An easier way is to use *copper tape* (Figure 7-10). Metallic copper is rolled very thin, backed with a tape adhesive, and placed on rolls. If you cut some lengths to firmly touch the wires, you have seven copper keys to use. Remember to reset your board at this point, as Circuit Playground Express will need to recalibrate to the new copper pads.

FIGURE 7-10. Copper tape makes excellent capacitive touch pads; use scissors to cut to fit.

Copper tape can also be used as circuit wires and for other types of electrical crafting. It does not flex well, so when using it in an application that bends or moves a great deal (like wearable electronics), consider using wires or conductive thread.

The Music Machine

Having demonstrated a piano-style "musical instrument," a more creative project is at hand. Not all instruments have to have piano-like keys. This is where people have been very creative.

For this next project, gather the following materials:

* Circuit Playground Express

* Alligator clip to wire end wires, Adafruit #3255 or similar

* 7 pieces of fruit (any kind: apples, limes, lemons, bananas, etc.)

Clip the alligator ends of the wires to the Circuit Playground Express pads marked A1, A2, A3, A4, A5, A6, and A7.

Press the other end of each wire into one piece of fruit—one fruit to one wire, no doubling up (Figure 7-11).

FIGURE 7-11. Inserting the wire into a piece of fruit. The lime does not feel a thing, honest.

You should now have a setup that looks similar to Figure 7-12 or Figure 7-13.

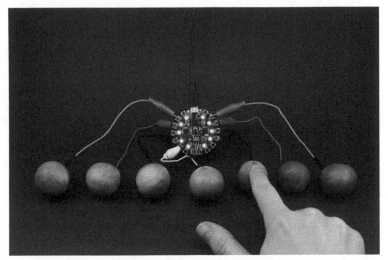

FIGURE 7-12. A fruit music machine using all limes. Photo credit: Kattni Rembor for Adafruit

FIGURE 7-13. Using different fruit types. Photo credit: John Park for Adafruit

The previous code for touch music works okay but does not have the best sound possible. Running the code in Example 7-1, you'll have a much more hip musical instrument.

The code in Example 7-1 also has some visual cues incorporated—when something is touched, a message is printed to the serial port and the 10 NeoPixels will light in different colors.

Example 7-1. The improved touch music code

```
# Fruit touch music synthesizer for Circuit Playground Express
# Original by Kattni Rembor for Adafruit Industries
from adafruit_circuitplayground.express import cpx
while True:
    if cpx.switch:
        print("Slide switch off!")
        cpx.pixels.fill((0, 0, 0))
        cpx.stop_tone()
        continue
    if cpx.touch_A4:
        print('Touched A4!')
        cpx.pixels.fill((15, 0, 0))
        cpx.start_tone(262)
    elif cpx.touch_A5:
        print('Touched A5!')
        cpx.pixels.fill((15, 5, 0))
        cpx.start_tone(294)
    elif cpx.touch_A6:
        print('Touched A6!')
        cpx.pixels.fill((15, 15, 0))
        cpx.start_tone(330)
```

```
elif cpx.touch_A7:
    print('Touched A7!')
    cpx.pixels.fill((0, 15, 0))
    cpx.start_tone(349)
elif cpx.touch_A1:
    print('Touched A1!')
    cpx.pixels.fill((0, 15, 15))
    cpx.start_tone(392)
elif cpx.touch_A2 and not cpx.touch_A3:
    print('Touched A2!')
    cpx.pixels.fill((0, 0, 15))
    cpx.start_tone(440)
elif cpx.touch_A3 and not cpx.touch_A2:
    print('Touched A3!')
    cpx.pixels.fill((5, 0, 15))
    cpx.start_tone(494)
elif cpx.touch_A2 and cpx.touch_A3:
    print('Touched A2 and A3 at the same time!')
    cpx.pixels.fill((15, 0, 15))
    cpx.start_tone(523)
else:
    cpx.pixels.fill((0, 0, 0))
    cpx.stop_tone()
```

The new code will play a musical note for as long as you touch the pad (fruit). When you are not touching anything, the `cpx.stop_tone` function at the bottom turns the sound off.

Also, if you touch the fruits connected to A2 and A3 at the same time, it will make a new tone, giving you eight keys (which can make a full octave).

If you have no reaction, move the slide switch. The slide switch makes the project not light up or play tones even when the fruit is touched. This is to provide some peace and quiet at home or in a classroom when you are not playing the instrument. Note the `cpx.pixels.fill`; it takes a tuple of three numbers for the red, green, and blue values for the NeoPixels.

If the touch is not very sensitive, use the following function after importing the libraries. You may have to adjust the value to be higher or lower than 50 for best results with your setup.

```
cpx.adjust_touch_threshold(50) # Set to be fairly sensitive
```

This same code will work without the fruit as done earlier, either with wires in a piano layout or just by touching the pads. Feel free to try it out.

Using Sound Files

At this point, the code in MakeCode and CircuitPython has focused on making individual musical notes. The CircuitPython library for Circuit Playground Express can also play sounds encoded in WAV format with specific encoding parameters.

The syntax for using the library is

```
from adafruit_circuitplayground.express import cpx
cpx.play_file("laugh.wav")
```

The file must be loaded onto the Circuit Playground Express flash drive. In the previous code, the file laugh.wav would be in the top-level (root) directory. If you put sounds in subdirectories, specify the subdirectory name before the filename:

```
cpx.play_file("/sounds/laugh.wav")
```

> NOTE WAV files must be encoded at 22,050kHz or less, 16-bit, and in mono (not stereo). For assistance in determining if the sound files you have meet these requirements, see a recent guide on the Adafruit website at *https://learn.adafruit.com/microcontroller-compatible-audio-file-conversion*.

The following program (Example 7-2) is a modification of the touch program listed earlier. The musical notes are replaced with seven WAV files. The samples used in this program are of the correct encoding and are freely available in a ZIP format file at *http://adafru.it/fruitBoxSamples* (scroll down to the fruitBoxSamples.zip button to download).

Example 7-2. The touch WAV sound program

```
# Touch wave file player for Circuit Playground Express
#       The .wav files can be downloaded
#       from adafru.it/fruitBoxSamples
from adafruit_circuitplayground.express import cpx
while True:
    if cpx.switch:
        print("Slide switch off!")
        cpx.pixels.fill((0, 0, 0))
        cpx.stop_tone()
        continue
    if cpx.touch_A4:
        print('Touched A4!')
        cpx.pixels.fill((15, 0, 0))
        cpx.play_file("fB_elec_blip2.wav")
    elif cpx.touch_A5:
        print('Touched A5!')
        cpx.pixels.fill((15, 5, 0))
        cpx.play_file("fB_bd_zome.wav")
    elif cpx.touch_A6:
        print('Touched A6!')
        cpx.pixels.fill((15, 15, 0))
        cpx.play_file("fB_bass_hit_c.wav")
    elif cpx.touch_A7:
        print('Touched A7!')
        cpx.pixels.fill((0, 15, 0))
        cpx.play_file("fB_drum_cowbell.wav")
    elif cpx.touch_A1:
        print('Touched A1!')
        cpx.pixels.fill((0, 15, 15))
        cpx.play_file("fB_bd_tek.wav")
    elif cpx.touch_A2:
        print('Touched A2!')
        cpx.pixels.fill((0, 0, 15))
        cpx.play_file("fB_elec_hi_snare.wav")
    elif cpx.touch_A3:
        print('Touched A3!')
        cpx.pixels.fill((5, 0, 15))
        cpx.play_file("fB_elec_cymbal.wav")
    else:
        cpx.pixels.fill((0, 0, 0))
        cpx.stop_tone()
```

The code plays one of seven WAV files depending on which touchpad is touched. The slide switch silences the sounds. If you have no sound, check the slide switch.

The WAV files can be recordings of anything. The BBC in the United Kingdom has a free noncommercial library they released in 2018 full of sounds. You'll find it at *http://bbcsfx.acropolis.org.uk/*.

The freeware program Audacity can record WAV files or convert files from one format to another. The WAV files should be small, with a 22,050 kHz sample rate (or lower), 16-bit PCM, and monophonic. You really cannot put full songs on Circuit Playground Express. The sound is good but it's not like an MP3 player, and songs in WAV format are very large.

Your Turn: Capacitive Touch

If you would like to work on capacitive touch in programs, try the following changes and answer the following questions:

* At the end of the music program, extra if statements were added to make a new tone if both A2 and A3 were touched at the same time. You can extend the program to add tones for when other combinations of pads are touched together— maybe add a new tone if three pads are touched together or a push button is used?

* What other NeoPixel colors can be used when a pad or pads are touched? Remember, the values for each color are tuples of red, green, blue and each value can range from 0 to 255. An online color picker is available at *www.w3schools.com/colors/ colors_rgb.asp*.

* Try out different WAV sound files on your Circuit Playground Express.

EMULATING A COMPUTER USB KEYBOARD

An excellent example of applying Circuit Playground Express to real-world uses is programming the board to act as a keyboard and/or a computer mouse. The USB interface standard has specific, standard protocols for human interface devices (HIDs). HIDs include keyboards, mice, MIDI musical codes, and more. The microcontroller on Circuit Playground Express has the necessary capability to control the USB port to emulate an HID device. The following example will demonstrate this application.

For this application, an additional library is needed to implement all the hardware-level communication on USB. Fortunately, Adafruit provides an open source library for implementing HID on Circuit Playground Express. The `adafruit_hid` library is available via GitHub at *https://github.com/adafruit/ Adafruit_CircuitPython_HID*.

USB Port Recognition

In normal CircuitPython mode, Circuit Playground Express will emulate a flash drive called CIRCUITPY when plugged into USB.

When you program the board to act as a USB keyboard or mouse, the board should work both as a flash drive named CIRCUITPY and as a keyboard/mouse at the same time.

If you have issues, unplugging Circuit Playground Express and plugging it back in appears to work fine to get both functions working.

To remove keyboard and/or mouse functionality, rename `code.py` on the drive and press Reset. You may also load another CircuitPython program as `code.py` on the board or delete the `code.py` program file.

Table 7-1 is a quick overview of the key functions available in the HID library. First, there are press and release key functions. To simulate pressing the A key, that key would first programmatically be pressed, then released. Those functions are done by the `press` and `release_all` functions. To combine a press and a release, the `send` function performs a `press` and a `release_all`.

TABLE 7-1. Library functions in the `adafruit_hid` library

LIBRARY FUNCTION	CIRCUITPYTHON CODE
Defining a keyboard object for use	`from adafruit_hid.keyboard import Keyboard` `kbd = Keyboard()`

Defining individual keys, including Shift and Ctrl	`from adafruit_hid.keycode import Keycode`
	`kbd = Keyboard()`
	
	`kbd.press(Keycode.CONTROL, Keycode.X)`
	`kbd.release_all()`
	`kbd.press(Keycode.RIGHT_ARROW)`
	`kbd.release_all()`
	`kbd.send(Keycode.SHIFT, Keycode.A)`
	First are modifiers (if needed) such as Ctrl, Shift, Alt, then the key itself. The send function combines the press and release_all functions.
	For a list of key codes, see *https://github.com/adafruit/Adafruit_CircuitPython_HID/blob/master/adafruit_hid/keycode.py*.

Defining keyboard layout and sending strings of text	`from adafruit_hid.keycode import Keycode`
	`from adafruit_hid.keyboard_ layout_us import KeyboardLayoutUS`
	
	`kbd = Keyboard()`
	`layout = KeyboardLayoutUS(kbd)`
	
	`# Type 'abc' then Enter (a newline).`
	`layout.write('abc\n')`

LIBRARY FUNCTION	CIRCUITPYTHON CODE

Retrieve the necessary key codes for a particular character	```python
	from adafruit_hid.keyboard import Keyboard
	from adafruit_hid.keyboard_layout_us import KeyboardLayoutUS
	kbd = Keyboard()
	layout = KeyboardLayoutUS(kbd)
	# Get the keycodes to type a '$'.
	# The method will return (Keycode.SHIFT, Keycode.FOUR).
	keycodes = layout.keycodes('$')
	``

Press a key and release a key	```python
	from adafruit_hid.keyboard import Keyboard
	from adafruit_hid.keycode import Keycode
	# Set up a keyboard device.
	kbd = Keyboard()
	# Press and hold the shifted '1' key to get '!' (exclamation mark).
	kbd.press(Keycode.SHIFT, Keycode.ONE)
	# Release the ONE key
	kbd.release(Keycode.ONE)
	``
	Note: Most often it is easier to code to use the Release All Keys function.

Release all keys	```python
	from adafruit_hid.keyboard import Keyboard
	from adafruit_hid.keycode import Keycode
	
	# Set up a keyboard device
	kbd = Keyboard()
	
	# Type control-x
	kbd.press(Keycode.CONTROL, Keycode.X)
	kbd.release_all()
	
	# Type capital 'A'
	kbd.press(Keycode.SHIFT, Keycode.A)
	kbd.release_all()
	``

HID Is a Frozen Library

The HID library is always available to CircuitPython programs. As of CircuitPython version 2.3.0, HID is a frozen library. A frozen library is part of the core of CircuitPython as of version 2.3.0 and higher. No file in the Circuit Playground Express /lib directory is required to support the HID library.

If you have upgraded from a CircuitPython version prior to 2.3.0 and you have library files in the /lib subfolder on the CIRCUITPY drive, you will want to delete the HID library in /lib. You do this by deleting the directory and files in /lib/adafruit_hid/. This will ensure you are using the frozen version of the HID library.

The online documentation is also available at *https:// circuitpython.readthedocs.io/projects/hid/en/latest/*.

Support for Non-US Keyboards Is Lacking

As of this book's publication, the adafruit_hid library only provides the keyboard layout KeyboardLayoutUS. More KeyboardLayout classes may be added to handle non-US keyboards and the different input methods provided by various operating systems.

If the host is expecting a non-US keyboard, the character to key mapping provided by KeyboardLayoutUS will not always be correct. Different keypresses will be needed in some cases. For instance, to type an A on a French keyboard (AZERTY instead of QWERTY), Keycode.Q should be used.

Circuit Playground Express's two onboard buttons will be used to act like two keys on a USB keyboard. You, the programmer, are not limited to emulating the 104 or so keys on a standard keyboard. It would be unproductive having the left button emulate an A key and the right a B key, for example. There are some interesting multiple key combinations that may be helpful. In addition, you can program multiple keys (to spell out a long series of text characters) with one button press.

The following are listings for CircuitPython programs to perform interesting keyboard emulation tasks. The Alt-Tab functionality works for Microsoft Windows and Chrome OS. Ctrl-K appears to work in Google Docs and perhaps other applications. Feel free to change the code to functions your computer will recognize.

```
import time
from adafruit_hid.keyboard import Keyboard
from adafruit_hid.keycode import Keycode
from adafruit_circuitplayground.express import cpx
kbd = Keyboard() # Create a keyboard object to work with
while True:
    # press ALT+TAB to swap windows
    if cpx.button_a:
        kbd.press(Keycode.ALT, Keycode.TAB)
        kbd.release_all()
    # press CTRL+K, which in a web browser will open
    #    the search dialog
    elif cpx.button_b:
        kbd.press(Keycode.CONTROL, Keycode.K)
        kbd.release_all()
    time.sleep(0.1)
```

Here is another CircuitPython program that outputs text:

```
# Circuit Playground Express Keyboard Emulation Example
# 2018 for Getting Started with Circuit Playground Express
import time
from adafruit_hid.keyboard import Keyboard
from adafruit_hid.keyboard_layout_us import KeyboardLayoutUS
from adafruit_circuitplayground.express import cpx
kbd = Keyboard()
```

```
layout = KeyboardLayoutUS(kbd)
while True:
    if cpx.button_a:
        # Type your name followed by Enter (a newline).
        layout.write('Kim Possible\n')
        # text should not be too long
    time.sleep(0.1)
```

Type the program into your editor, such as Mu or Caret. Check for mistakes.

Plug your Circuit Playground Express board into the USB cable connected to your PC. When the USB flash drive CIRCUITPY appears, download your code to the drive with the file named `code.py`.

Unplug your board and plug it back in. Circuit Playground Express should now be running as a USB keyboard. You can verify this by going to the operating system device manager and looking for a new device (see Figure 7-14).

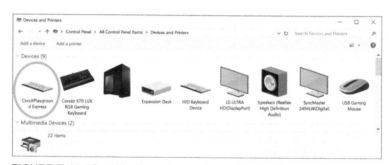

FIGURE 7-14. The Windows Control Panel → Devices and Printers shows a new keyboard named Circuit Playground Express (the first device, circled in blue).

Open up a text editor or word processor. Press the A and B keys to see what they do. Look up other key codes, edit the program, and try your codes.

For the second demonstration, the CPX and HID libraries are very large. If you put in much text for the button to activate, a memory error may occur. In that case, first ensure you are using CircuitPython 2.3.0 or later and that you are using the frozen version of the CPX library (see the "Using the Internal (Frozen) CPX Library" sidebar in Chapter 6). If you still experience issues, go to *learn.adafruit.com* and search for the CircuitPython Essentials guide for performing low-level button recognition coding. Using low-level libraries will save more memory than using the CPX library but at the cost of greater code complexity.

Your Turn: Keyboard Emulation

Change the code to

* Try out different types of keypresses or strings.

* Advanced: Use the capacitive touchpads as extra keys and program so that when the pads are touched, selected USB keypresses are performed.

MOUSE EMULATION

The Adafruit HID library for CircuitPython also has functions for performing mouse functions such as clicking the mouse buttons, using the scroll wheel, and moving the mouse cursor.

Here are the functions for mouse emulation. At the beginning of your Python code, you should import the Mouse library:

```
from adafruit_hid.mouse import Mouse
m = Mouse()
```

The functions available are shown in Table 7-2.

TABLE 7-2. Adafruit HID functions for mouse emulation

LIBRARY FUNCTION	CIRCUITPYTHON CODE

Click mouse button(s)	`m.click(Mouse.LEFT_BUTTON)`
	`m.click(Mouse.RIGHT_BUTTON)`
	`m.click(Mouse.MIDDLE_BUTTON)`
	`# call below does both buttons`
	`m.click(Mouse.LEFT_BUTTON |`
	`Mouse.RIGHT_BUTTON)`

Press and hold mouse button(s)	`m.press(Mouse.LEFT_BUTTON)`
	`m.press(Mouse.RIGHT_BUTTON)`
	`m.press(Mouse.MIDDLE_BUTTON)`
	`m.press(Mouse.LEFT_BUTTON |`
	`Mouse.RIGHT_BUTTON)`

Release mouse button(s)	`m.release(Mouse.LEFT_BUTTON)`
	`m.release(Mouse.RIGHT_BUTTON)`
	`m.release(Mouse.MIDDLE_BUTTON)`
	`m.release(Mouse.LEFT_BUTTON |`
	`Mouse.RIGHT_BUTTON)`

Double-click left button	`m.click(Mouse.LEFT_BUTTON)` `m.click(Mouse.LEFT_BUTTON)`
Move the mouse cursor and scroll wheel	`move(x=0, y=0, wheel=0)` Parameters: <ul><li>x—Move the mouse along the x-axis. Negative is to the left; positive is to the right.</li><li>y—Move the mouse along the y-axis. Negative is upward on the display; positive is downward.</li><li>wheel—Rotate the wheel this amount. Negative is toward the user; positive is away from the user. The scrolling effect depends on the host.</li></ul>Values should be from –127 to 127 (this may be expanded in future code releases, so check the latest Adafruit API documentation).

LIBRARY FUNCTION	CIRCUITPYTHON CODE
| Move the mouse cursor and scroll wheel (continued) | ```
# Move 100 to the left. Do not move up and down.
# Do not roll the scroll wheel.
m.move(-100, 0, 0)
# Same, with keyword arguments.
m.move(x=-100)

# Move diagonally to the upper right.
m.move(50, 20)

# Same
m.move(x=50, y=-20)

# Roll the mouse wheel away from the user
m.move(wheel=1)
# Roll the mouse wheel towards the user
m.move(wheel=-1)
``` |

The following is a demonstration program for the mouse capability. The A and B buttons emulate the mouse left- and right-click. Touching pads A2, A3, A4, or A5 (the upper pads) scrolls up; touching A6, A7, or A1 scrolls down.

```
# Circuit Playground Express Mouse Emulation Example
# 2018 for Getting Started with Circuit Playground Express
from adafruit_hid.mouse import Mouse
from adafruit_circuitplayground.express import cpx
m = Mouse()
cpx.adjust_touch_threshold(50) # adjust sensitivity
while True:
    if cpx.button_a:
        m.click(Mouse.LEFT_BUTTON)
    elif cpx.button_b:
        m.click(Mouse.RIGHT_BUTTON)
    elif cpx.touch_A2 or cpx.touch_A3 or cpx.touch_A4 or cpx.
touch_A5
        m.move(wheel=+1)
    elif cpx.touch_A6 or cpx.touch_A7 or cpx.touch_A1
        m.move(wheel=-1)
```
```

Although mouse emulation can be useful, in general it is often limited to mouse clicks in applications such as assistive technology where someone may not be able to use standard mouse buttons.

One important application using the mouse movement feature is *mouse jiggling*—the emulation of mouse movement every few seconds to keep screen savers and logout programs from terminating a running program.

Your Turn: Mouse Emulation

Change the code to make your own custom mouse commands:

* How would you code a "mouse jiggler" program?

Project Ideas: CircuitPython

Here are several ideas on taking this chapter's concepts to more advanced projects:

* Data logging: With serial communications, the data from sensors onboard the Circuit Playground Express and others, perhaps connected via the electronic pads, can be displayed and recorded. What type of applications would benefit from the data collected by the board? Think about all the sensors mentioned earlier in the book and think creatively.

* Assistive technology: Many people cannot operate a standard keyboard or mouse. Others may have difficulty pushing a standard push button. How can the Circuit Playground Express capability to translate keypresses, even capacitive touch, to other actions be helpful in assistive technology?

* With data coming out of the USB serial port, a computer-based program could look for specific text from Circuit Playground Express. Messages such as "Thirsty," "Light Needed," "Earthquake!," "Intruder!," or "Help!" could be output by the board and read on the PC for action. Describe ways in which Circuit Playground Express might provide an alert for action. Draw a diagram of how this might work.

WRAP-UP

This chapter provided some information and examples for creating CircuitPython programs using the onboard capabilities of Circuit Playground Express. It is a topic that could fill an entire book itself.

I suggest you continue your exploration in three ways:

* Continue to explore the CircuitPython libraries available. It is fun and rather easy to combine sensor input functions with output functions to realize many different and interesting projects.

* Look at the CircuitPython projects on the Adafruit Learning System at *https://learn.adafruit.com/category/express*. The CircuitPython guides provide many new CircuitPython programming methodologies you might wish to place into your own projects.

* Many people publish CircuitPython projects on the Internet—check out what others are doing.

CHAPTER QUESTIONS

1. Think of an example using code from this chapter and Circuit Playground Express to make a device to assist someone with limited mobility, who might have a hard time using a standard keyboard or mouse. Describe your idea.

2. The boot.py file runs CircuitPython when the board reboots or is powered on. This could be useful for running some code besides the code in code.py. Describe what CircuitPython code you might put in boot.py.

3. What Python statements provide for triggering certain code if an error is detected in the program?

8

Using the Arduino Development Environment

The popularity of the Arduino ecosystem for programming electronics has been growing for the last 10 years. Arduino grew from people who believed that it should be easy (and free) to program microcontrollers, especially for nontechnical Makers. The standard hardware built by the Arduino team provided inspiration for a variety of people, from hobbyists to fashion designers.

As the use of Arduino grew, so did both the company's software and hardware support. On the hardware side, Arduino issued more development boards with differing capabilities, expanding the number of people interested in using their hardware. Other companies joined in developing hardware and using open source programming software to allow the same easy programming experience that Arduino enjoyed. On the software side, useful hardware attracted talented programmers who developed excellent software. Similar to Python libraries, Arduino libraries were developed by the community growing around the Arduino hardware. The software support is, arguably, the greatest asset in the Arduino world today. Due to a large number of software developers, it is now possible to program many different microcontrollers with nearly the same Arduino code.

The Arduino software consists of two parts. The first is a microcontroller-specific *bootloader*, which is software on the microcontroller chip that communicates with a user's PC. Some call this software *firmware* to differentiate it from user-written code.

The other software is on the user's PC. It is called the Arduino integrated development environment (IDE). There are two versions: one that is downloaded software for specific types of operating systems (Windows, Mac, Linux) and a newer one, *Arduino Create*, which is web based.

The Arduino IDE is similar to Microsoft MakeCode in that it presents users with a software development environment where they can create code and download that code to the microcontroller (Figure 8-1).

Unfortunately, the newer Arduino environment for the web, Arduino Create, does not support Circuit Playground Express.

If you are in Chrome, navigate to the Chrome web store, *https://chrome.google.com/webstore/*. Search for "Arduino Create" and you will find the application shown in Figure 8-2. The app costs 99 cents per month.

FIGURE 8-1. The Arduino IDE software

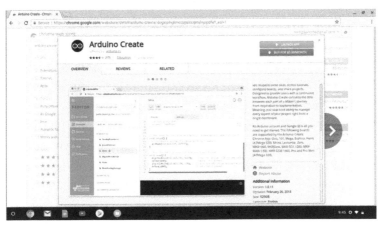

FIGURE 8-2. Arduino Create, a new Arduino environment for Chromebooks, but it does not support Circuit Playground Express and has a monthly fee.

This chapter provides the basics for getting started using the Arduino IDE to create programs for Circuit Playground Express.

THE ARDUINO PROGRAMMING LANGUAGE

The Arduino language is very much like the C programming language—but not exactly. Similar to C, it has added libraries to allow users to easily program the hardware in a microcontroller project. The hardware-specific code is in a library, so the syntax (the particular functions used by the user) stays the same if the type of microcontroller changes.

Also, more advanced software features are available in Arduino but are not commonly used. This includes programming in C++, an object-oriented version of C. But the Arduino developers encourage use of C programming to allow others to easily understand Arduino IDE programs.

Arduino programs are called *sketches* to distinguish them from other types of programs. Arduino sketches have the file extension .ino. Library files take after standard C programs and have the extension .h, which refers to a header file.

Note that learning Arduino is a bit less intuitive than either MakeCode or Python. If you want to learn to program quickly, start with MakeCode first. If you are a more experienced programmer and want more control, Arduino is great!

Why would users want to code their Circuit Playground Express using Arduino when MakeCode and CircuitPython are available? Arduino provides capabilities not available in the other languages:

* Library support: Arduino has an extensive set of libraries, and both internal and additional libraries are provided as open source by groups worldwide.

* Hardware support: Arduino code likely exists to interface a microcontroller to other hardware. For instance, most Adafruit hardware has examples coded in Arduino to assist their users. The company is working to provide more CircuitPython interface code, but it lags behind Arduino code.

* Memory: As C was developed to use code memory very efficiently to run on early computers, today the Arduino IDE–produced code is smaller and more memory efficient than in other languages. A complex CircuitPython program may use all of the Circuit Playground Express memory, whereas an equivalent Arduino program might take only a small fraction of the memory, leaving additional space for more functionality.

* Speed: As the Arduino IDE compiles C code into machine binary code, the resulting code runs very fast.

> **NOTE** If you are in a classroom environment, the Arduino environment is most likely already configured on your computer. If advised by your instructor, you can skip the installation and setup sections. You may want to follow the instructions later if you wish to configure an Arduino development environment for your own computer.

INSTALLING THE ARDUINO IDE

This section will guide you in downloading and installing the Arduino software on your computer. As of this writing, the Arduino Create web-based development environment does not provide the same breadth of hardware support as the software version. The one benefit of the web version is Chrome OS support.

The Arduino official website is at *www.arduino.cc*. If you search for "Arduino" on the web, you may find many results that provide Arduino information, but arduino.cc is the official website. On the site, you'll find many resources for learning to use Arduino to program projects. It is worth your time to browse the site for information on using the Arduino IDE and writing Arduino code.

Downloading the Software

In a web browser on your PC, navigate to *www.arduino.cc/en/ Main/Software*. Download the software for your operating system (Figure 8-3).

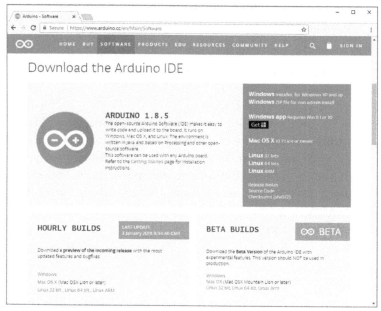

FIGURE 8-3. Downloading the Arduino software from the arduino.cc website

Install the software according to your operating system protocol. The installer should put an icon on your desktop (which is a teal circle with the Arduino logo inside).

The program is rather large compared to similar programs. This is due to using different open source software to compile a sketch into different hardware architectures. One of the benefits of Arduino is the breadth of different hardware supported.

Configuring the Arduino IDE

Circuit Playground Express is natively supported in the Arduino IDE, so it is easy to set up compared to other Arduino-compatible boards.

Start the software by double-clicking the Arduino icon. Then, open the Boards Manager by navigating to the Tools → Board menu and selecting Boards Manager at the top of the list (Figure 8-4).

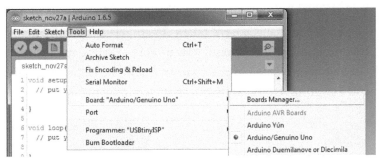

FIGURE 8-4. The Boards Manager menu in Arduino

A screen appears with additional software packages for the IDE. They are on the computer but not installed by default.

If you do not see "Arduino SAMD Boards (32-bits ARM Cortex-M0+)," you can type **Arduino SAMD** in the top search bar and look for the SAMD entry. On the right, ensure the version is 1.6.16 or later (the latest software is usually the best selection). Then click the Install button (Figure 8-5). Configuring the additional software will take a minute or two.

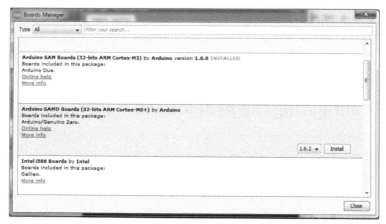

FIGURE 8-5. Select the SAMD Boards. Select the latest version, then click the Install button.

I recommend that you exit the Arduino software and rerun it to ensure the changes have taken effect. Select Quit and reopen the Arduino IDE. You should now be able to select and upload to the new boards listed in the Tools → Board menu. Select the Circuit Playground Express board, as shown in Figure 8-6.

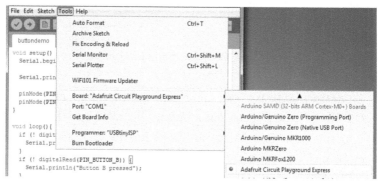

FIGURE 8-6. Selecting Adafruit Circuit Playground Express from the Tools → Board menu

Note that on Windows 10, operating support for communicating with many Adafruit boards is built in. If you have Windows 7, you will need a driver installed for the PC to recognize Circuit Playground Express.

Install Arduino Drivers (Windows 7 Only)

When you plug in the board, you may have to install a driver. First, go to *https://github.com/adafruit/Adafruit_Windows_Drivers/* and download the 32-bit (x86) or 64-bit (x64) depending on your version of Windows. Select which drivers you want to install. If you will be using other Adafruit products, you should consider installing all of the drivers listed, so you don't have to install additional ones later.

For an in-depth how-to, a tutorial with screen shots on the driver installation is available at *https://learn.adafruit.com/adafruit-circuit-playground-express/adafruit2-windows-driver-installation*.

Selecting the Serial Port

Plug Circuit Playground Express into your computer with a known good USB cable. Wait for the board to be recognized by the operating system (it just takes a few seconds). The operating system of the computer will create a serial/COM port to provide communication between the computer and Circuit Playground Express. Select the value with the label Circuit Playground Express (Figure 8-7).

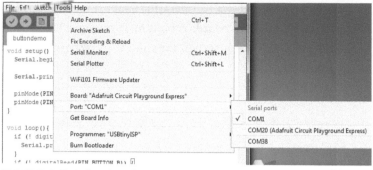

FIGURE 8-7. Selecting the Circuit Playground Express COM port

If you do not see a communications port labeled Circuit Playground Express as in Figure 8-7:

1. Ensure that you have connected the micro-USB end of the cable correctly into the Circuit Playground Express silver USB port (not the black plug-in opposite).

2. Ensure that the other end of the USB cable is connected to a USB port on your computer.

3. You should see the green power light if it's plugged in correctly; if not, check again.

4. Try again to see if the computer sees the board by clicking the Tools → Port again (even if it was on the screen).

   If it's not there still, try the next steps.

5. If your computer has multiple USB ports, try a different physical USB port. At times, the operating system might not detect Circuit Playground Express on a specific port, especially if the port is on a USB hub.

6. Check again via Tools → Port.

   If it's still not selectable in the menu:

7. Try another USB cable. Even if Circuit Playground Express powered up (green light), some cables (like for phones)

might be "charge only" and lack the data wires necessary for communications.

When you are able to select the correct communications port, the Arduino IDE should be ready for programming Circuit Playground Express.

STRUCTURE OF AN ARDUINO PROGRAM

Arduino code has two main sections. The first is in a function called setup. This routine is run once at the beginning of the program. It is equivalent to the on start block in MakeCode, or the statements that come before a while True loop in CircuitPython. The second function used in Arduino code is called loop. It is exactly like the MakeCode forever loop and the CircuitPython while True loop. Any code in the loop function will run over and over again. It's where you put code to constantly monitor a project and/or provide interactivity.

When you open a new Arduino program in the Arduino IDE by selecting File → New, the following program template will automatically be filled in:

```
void setup() {
// put your setup code here, to run once:
}
void loop() {
// put your main code here, to run repeatedly:
}
```

At this point, you can start typing in code or pasting it from another source such as a code editor. We'll discuss a few more basics that would be useful for you to learn at this point before typing in a program.

Like CircuitPython (but not MakeCode), Arduino uses libraries to bring specific hardware and specialized software functionality to a program. The Arduino team has implemented a default set of libraries for use on typical hardware. They are documented on the arduino.cc site under Reference. Also, thousands of third-party Arduino-compatible libraries are available for use. A majority of third-party libraries are open source, which means they are free to use (but usually require attribution of the original author).

The typical Arduino code looks like this:

```
/* The documentation at the top of the program starts with a
forward
slash followed by a star. All text will be treated as
comments until ended by a star followed by a forward slash.
*/
/* Declare libraries as the first thing in a program */
#include <arduino.h>
#include <anotherlibrary.h>
/* This is where you declare variables used by all functions
*/
/* 3 different variables are below, c is initialized to 0 */
int a;
long b;
float c = 0.0;
/* All non-function statements in C end in a semicolon */
/* functions and statements like if enclose the contents in
{ } */
void setup() { /* place any setup code in this function */
    if( c == 0.0 ) {
        a = 1;
        b = 2;
    }
}
void loop() {
/* code is executed forever in a loop placed here */
/* The variable d is only used in this loop */
    int d;
    d = a + b + c;
}
```

Looking at the structure of the sketch, several *comments* are placed in the code. They start with /* and end in */. A comment can span multiple text lines (as at the beginning of the previous code) or just part of a single line. Single-line comments may also be started with two forward slash characters (//).

You can include libraries in your code by listing them in a C `include` statement at the start of the code. The syntax is a hashtag sign (#) immediately followed by the word `include`, then a library name enclosed within less than (<) and greater than (>) symbols. Library names are identical to the filename of the library on your computer and have an .h file extension that labels the file as a header file. The `arduino.h` library contains all hardware interaction functions for Arduino code and is usually included first. Some coders do not include `arduino.h` (or forget to do so); the code will not show an error because the Arduino IDE will automatically include the library if it is missing, but it is a best practice to explicitly list `#include <arduino.h>` as the first actual (non-comment) line in the code.

Any other libraries are "included" after `arduino.h`. The sketch example lists a library on the user's computer named `anotherlibrary.h`. There is no limit to the number of libraries that may be used in a sketch. But a good practice is to not include a library you do not intend to use. Why? The unused code in the library will bloat the sketch. Many microcontrollers, like Circuit Playground Express, have limited flash and program memory. Free memory is memory that can be used for more functionality later in development.

In Arduino, you must declare—that is, explicitly name—each variable before you use it. The default availability (called *scope*) of a variable in C is within the block where you *declare* and use it. Variable names can be upper- and lowercase letters, numbers, and the underscore (_) symbol. By convention, most variable names are lowercase.

The variable type is a new concept. You must explicitly declare what type of variable will be used in the program. Here's a list of variable types:

* `int` (integer)—A value without a decimal point. Typical ranges for an integer are –32,768 to 0 to 32,767. Examples are 279, 1001, 0, –23, –990.

* long—A large integer; can be a value from –2,147,483,648 to 2,147,483,647.

* float (floating-point numbers)—Numbers with a decimal point and a fractional amount. Examples are 3.1415, –22.2, 0.0, 430000.01. Numbers can be as large as 3×10 to the 38th power.

* char (a single character)—For example, reading serial data may involve a receive function providing a character value when data is received. A character may be any symbol on the keyboard (0 to 255).

* A unsigned short integer, uint8_t, is often used by functions; it also ranges from 0 to 255.

So when a variable is to be used in a program, you declare the variable by stating its type, stating the variable name, and finishing with an end-of-statement semicolon. Optionally in a declaration, you may set the initial value. In the previous example, the variable a is declared as an int. The variable b is declared as a long (large integer). Finally, the variable c is declared as a float decimal number and *initialized* (set at start) to 0.0. Like all single-line lines of code in C, the declaration ends in a final semicolon (;) character.

The three example variables are declared before the setup and loop functions. If variables are declared between the include statements and the setup function, they are considered global variables, which means they are usable by any function the user writes in the program (but not in library code). So a global variable may be changed by setup and/or loop and/or any other function the user creates.

Now look at the loop function. We declare an additional int variable named d. This variable can be used only in the loop function. setup or any other functions will not see it. Using global variables can help simplify variable use, but note that if there was a global variable d and a local variable declaration named d, that function will use the local variable instead of the global variable, and that could lead to confusion. Local function variables are often used for loop counters or intermediate math results.

In the Arduino IDE, having to declare each variable can be tedious and complicated when compared with Python. Why? C is a *strongly typed language* in comparison to CircuitPython. Every variable must be declared and placed where it needs to be, depending on how it will be used. Python variables, on the other hand, are just another object that the program manipulates. If the program encounters a new object, it deals with it. Languages that are not strongly typed can be easy to code, but they do not allow the fine-tuning that C and other strongly typed languages have. Most often, the hidden firmware code that is between the user program and the hardware is coded in C and compiled into native hardware machine language. This is due to the small code and data sizes of C code, which does not require the overhead of object manipulation of Python.

Compared to variables, the function declarations are fairly straightforward:

* The Arduino functions `setup` and `loop` are the same as the MakeCode `on start` and `forever` loop blocks. `setup` will be executed once at the beginning of a program; `loop` will continue to run until Circuit Playground Express is reset or a new sketch is loaded.

* Functions in C/Arduino that do not return any value (and both `setup` and `loop` do not), are *declared* (or labeled) void to say there is no return value. It does not matter to us other than it should be placed before the name of the functions.

* After a function name is a list of variables to send into a function surrounded by parentheses. As `setup` and `loop` do not have function values, there is no text between the opening and closing parentheses.

* The *body* (code to be executed) of the function starts with a curly open brace, {, and ends with a curly closed brace, }. This is similar to the indented portion of a Python loop.

Code within a function is indented as in Python. But unlike Python, indentation in Arduino is optional (but highly recommended). The curly braces, { and }, enclose the function code as a block, so it is not the indentation that defines the block as in Python. Also, lines of code in if statements and other loops like while are also indented. Generally, several spaces are used for each loop or if statement (although some use more or less, and some use tabs). The Arduino IDE will not give an error due to differing indentation (or a lack of indentation). But anyone looking at code or maintaining it after being written will thank the original programmer if the code is indented consistently and commented well, and is easy to follow.

The best way to understand a language is to see an example. For Arduino, the standard example is a program that blinks a single LED light. The following code sets up the Circuit Playground Express D13 red LED and makes it blink:

```
/* Blink Sketch for Circuit Playground Express */
#include <arduino.h>
void setup() {
/* initialize digital pin 13 (the red LED) as an output */
    pinMode(13, OUTPUT);
}
// the loop function runs over and over again forever
void loop() {
    // turn the LED on (HIGH is the voltage level)
    digitalWrite(13, HIGH);
    delay(1000); // wait for a second
    // turn the LED off by making the voltage LOW
    digitalWrite(13, LOW);
    delay(1000); // wait for a second
}
```

The function `pinMode` takes two values: the pin number on the board (D13 is the small LED on a majority of Arduino-compatible boards) and a second value that sets the pin to be an OUTPUT. If you need to read a value from a Circuit Playground Express data pad, you could use the same function with the number of the pad and the value INPUT.

The function `digitalWrite` also takes two values: the pin number to work with and a value of HIGH or LOW. HIGH puts a voltage of 3.3 volts on the pin; LOW sets the voltage to 0. For pin D13, there is the LED plus a current limiting resistor connected to ground (0 volts). If the pin is switched to 3.3 volts (which is HIGH) by the program, electric current flows and the LED lights. If the program switches the value to LOW, there is no electrical current and the LED is dark.

The function `delay` takes an integer that is the number of milliseconds to wait (sleep). A value of `1000` is 1000 milliseconds, which is one second. Unlike in Python, the `delay` function is part of the Arduino library and so no reference to a time library is needed.

The knowledge needed to write an equivalent LED blink program is much higher for Arduino than for CircuitPython or MakeCode. You have to learn the syntax for the C programming language and a list of functions to perform actions on the hardware. The entire reference for Arduino functions is available at *www.arduino.cc/reference/en/*.

If you would like to learn more about the C programming language, there are many sites on the Internet to help you. I like *www.tutorialspoint.com/cprogramming/*—it is clear and has less technical asides than others. The classic book *The C Programming Language*, by Brian Kernighan and Dennis Ritchie, is excellent and holds up well.

> **NOTE** In C reference material, the main program is called `main`. In the Arduino programming language, the required functions are `setup` and `loop`. When the Arduino IDE compiles the code, it sticks in the required `main` for the C language, which calls `setup` once and calls `loop` in a `forever`-style loop. So when you read a C reference, keep this in mind. Also, not all C standard functions are defined in the Arduino reference language. Consult *www.arduino.cc/reference/en/* for details on which functions Arduino supports.

UPLOADING CODE TO CIRCUIT PLAYGROUND EXPRESS

The Arduino IDE directly programs Circuit Playground Express via the USB connection. This is why the IDE requires the serial information for the board's connection to the PC. It is not possible to copy a code file, such as a UF2 file, to the board for Arduino programming. It is not a terribly difficult process—just a bit different.

Open the Arduino IDE on your computer. Copy or type in the Blink sketch. Choose File → Save As and save your file as `blink.ino` (Figure 8-8). All Arduino sketches end in an `.ino` extension.

Next the code will be compiled (changed) from Arduino/C to binary code and loaded to Circuit Playground Express.

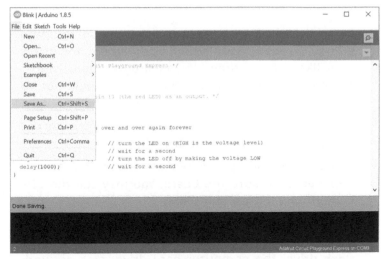

FIGURE 8-8. Choosing File → Save As and saving the sketch

Uploading a Sketch

Uploading an Arduino sketch is similar to loading a MakeCode program or a CircuitPython program. But there are some differences.

In CircuitPython, Circuit Playground Express acts like a flash drive connected to the programming computer. Arduino is closer to MakeCode in that the Arduino IDE puts the code on Circuit Playground Express and sets it running. There is no flash drive. And if you had previously set the board up for CircuitPython, that environment is wiped free.

Plug Circuit Playground Express into the computer USB port via a USB cable (again, a known good cable, please). To check if the board is recognized by the operating system, choose Tools → Port to see what device(s) the Arduino IDE recognizes.

> **WARNING** If you have been using the Circuit Playground Express CircuitPython environment, the board acts like a flash drive. If you program an Arduino sketch onto the board, the flash drive should not be erased. But this behavior is not guaranteed.
>
> If you have Python programs or other files on the board, save the files to another disk.
>
> The CircuitPython programming environment can be reloaded back onto the board. You may see the CircuitPython files are still there.
>
> Please keep a backup of all files from Circuit Playground Express on another drive, be it a dedicated flash drive, the computer hard drive, or a network storage device.

The connection of the USB cable will create a serial communications (COM) port. If you do not see a COM port for Circuit Playground Express under Tools → Port, check your cable and your computer connection.

At times, the computer may not detect that Circuit Playground Express is connected to the computer. In this instance, the board should be placed in *bootloader mode*. To get into bootloader mode, press the small Reset button on the board twice. All the circular NeoPixel LEDs should glow red momentarily, then a steady green. If the LEDs are not green, press Reset twice (not too fast or slow) to have them turn green (Figure 8-9). There is a sweet spot to the button pushes; it may take a couple of tries.

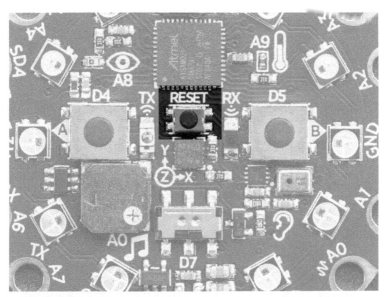

FIGURE 8-9. Press the Circuit Playground Express Reset button twice for bootloader mode.

You can now select the port corresponding to Circuit Playground Express from the Tools → Port menu. The serial port should be indicated by the IDE as a Circuit Playground Express board (Figure 8-10).

After the correct communications port is set, Circuit Playground Express should be communicating with the computer. At this point, the Arduino IDE converts the Arduino code to machine binary runnable code and loads it via serial—very different from CircuitPython.

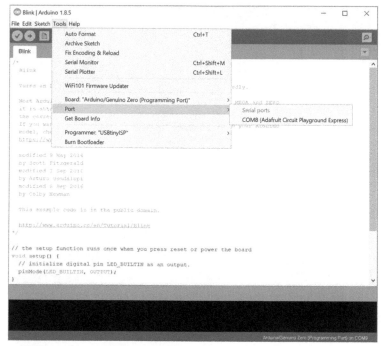

FIGURE 8-10. Selecting the serial COM port in the Arduino IDE Tools → Port menu

The process of taking the Arduino code and converting it to binary code that Circuit Playground Express can run directly is called *compiling*. With CircuitPython, the code is converted to binary on the board itself. With Arduino, the binary code is generated inside the Arduino IDE and then loaded onto the board (Figure 8-11).

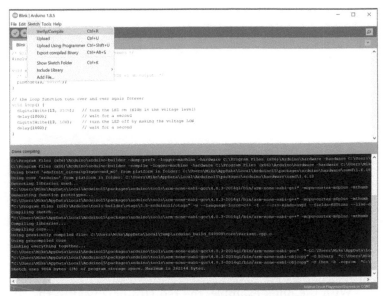

FIGURE 8-11. The code verification and compilation in the Arduino IDE Sketch menu

When you click Verify/Compile, the IDE will check the code to make sure it is 100 percent correct in terms of a valid C code with valid include files. If any *syntax errors* are detected, verification will stop with an error stating where the problem was found. If the code verifies correctly, the IDE will *compile* the code and generate the corresponding binary code. At the end of the process, the size of the binary file and the maximum code size for Circuit Playground Express will be displayed. Most programs will never reach the maximum code size, unlike older boards with smaller internal flash memory sizes.

To upload the file, select Sketch → Upload from the Arduino IDE menu (Figure 8-12). When you upload a file, any file on Circuit Playground Express will be replaced by the new binary file. The CIRCUITPY flash drive will not be shown until CircuitPython is reloaded onto the board.

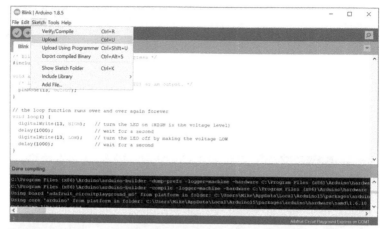

FIGURE 8-12. Uploading the compiled code to Circuit Playground Express

With the Blink example, the red D13 LED next to the USB connector on Circuit Playground Express should be blinking one second on, one second off.

The process of editing a program in the IDE and then using Sketch → Verify/Compile and Upload will become second nature the more you use the program. I cannot say the same for some errors you might encounter.

If the program will not verify, read the error message displayed in the black portion of the IDE. You might need to expand the IDE window size to see messages clearly. Check for the basics:

* All variables are declared either globally after the include statements or within a function as appropriate.

* You have semicolons where needed and enclose multiline statements like functions, if statements, etc. in curly braces, { }.

* You must have both a setup function and a loop function— they are not optional, although you do not need to have any code in them.

You can look up error messages in a web-based search engine to look for more esoteric causes.

THE CIRCUIT PLAYGROUND ARDUINO LIBRARY

As in CircuitPython, there are low-level functions like `digitalRead` and `digitalWrite` that can access both onboard sensors and communicate over the outer signal pads. Also, as in CircuitPython, Adafruit has an open source library where you can easily access sensor values and light NeoPixels without having to perform low-level function calls. Why would it be preferable to use the Circuit Playground library versus Arduino native pin functions?

* It takes significantly less time to use the library, allowing you to focus more on the functionality you want with a project.

* Some code, as for NeoPixels, is just a rather complicated set of signal timing code best left under the hood.

* Some sensors, like temperature and light, provide values that must be mathematically changed to get standard units. This is done by the library so that you do not have to type in the code from a reference sheet.

The coding adage is "If it's been done by someone else, consider using that and avoiding reinventing the wheel."

Note that the Arduino library combines support for the Adafruit Circuit Playground "classic" board (based on the Atmel 32u4 chip) and the Circuit Playground Express. The function calls noted will work with Express. Some of the examples in the library package are tailored for one or the other boards, so it's good to keep that in mind. If there are two similar examples, look for the one with Express in the name.

Installing the Circuit Playground Arduino Library

The Arduino IDE has a Library Manager built into the program. From the menu bar in the upper-left corner of the window, select Sketch → Include Library → Manage Libraries (Figure 8-13).

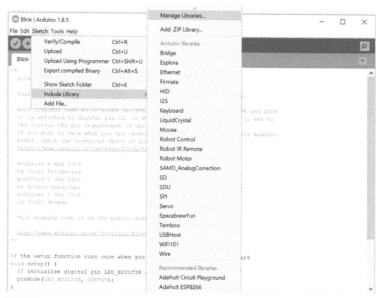

FIGURE 8-13. Choosing Manage Libraries

The window shown in Figure 8-14 opens. In the search box, type **adafruit** and press Enter to list Arduino libraries by Adafruit Industries. Scroll down the list until you get to Circuit Playground.

Click the Adafruit Circuit Playground library box. You will see a Select Version drop-down box and an Install button (Figure 8-15). Select the largest number for the version and click Install. If the library is already installed, you can click the Update button.

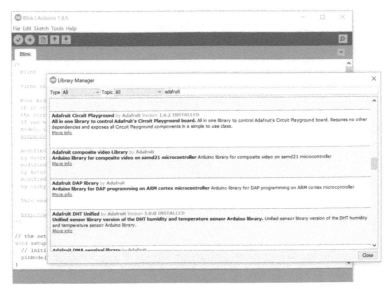

FIGURE 8-14. The Adafruit Circuit Playground library in the Arduino Library Manager

FIGURE 8-15. Installing and updating the Adafruit Circuit Playground Arduino library

In Figure 8-15, the computer already had version 1.6.2 of the library installed and version 1.8.0 was available. Clicking the Update button provided the latest software for using the Circuit Playground.

At this point, when you wish to use the Circuit Playground library (or any other installed library), choose Sketch → Include Library, and then select the library you want from the list (Figure 8-16).

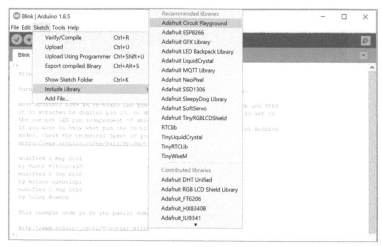

FIGURE 8-16. Selecting the Adafruit Circuit Playground library from the Sketch menu

Clicking the library name Adafruit Circuit Playground will put the following line in your code:

```
#include <Adafruit_CircuitPlayground.h>
```

If you need to relocate the `#include` statement below comments or other library include lines, that's fine. Just remember all include lines should come before any program code.

> NOTE If you would like to browse the code, view examples, note a bug, or contribute code, Circuit Playground library source code is located on GitHub at *https://github.com/adafruit/Adafruit_CircuitPlayground*.

CIRCUIT PLAYGROUND LIBRARY FUNCTIONS

Table 8-1 lists the functions available in the Circuit Playground Arduino library. Note that the Serial.println statements are examples of printing to the serial port (the console area at the bottom of the Arduino IDE window). They are nearly identical to the print statements in CircuitPython, but the Arduino function call must end in a semicolon like all C statements.

You can put any code you wish with the call; using a C if statement is just one way to test for the value.

TABLE 8-1. Adafruit Circuit Playground Arduino library functions

LIBRARY FUNCTION	ARDUINO CODE

begin	Necessary function to initialize library functions.
	
	``void setup() {
	    CircuitPlayground.begin();
	}```
	
	begin can contain an optional value, brightness, for the NeoPixel brightness.
	
	``void setup() {
	    CircuitPlayground.begin(50);
	}```

Test for Express	Test to see if the board is a Classic or an Express:
	
	``void setup() {
	    CircuitPlayground.begin();
	    if(!CircuitPlayground.isExpress()
	) {
	        Serial.Println("Board is not an
	Express!");
	        while(1) ; // infinite loop
	    }
	}```

LIBRARY FUNCTION	ARDUINO CODE
| Button A (left) | ```
if (CircuitPlayground.leftButton()) {
    Serial.println("Left button pressed!");
}
``` |
| Button B (right) | ```
if (CircuitPlayground.rightButton()) {
    Serial.println("Right button pressed!");
}
``` |
| Slide switch | ```
if (CircuitPlayground.slideSwitch()) {
    Serial.println("Slide to the left");
} else {
    Serial.println("Slide to the right");
}
```
(True to the left, False to the right) |
| Red (D13) LED | ```
CircuitPlayground.redLED(HIGH);
delay(100);

CircuitPlayground.redLED(LOW);
delay(100);
```
(HIGH turns LED on; LOW turns LED off) |
| Temperature sensor value | ```
float temp_c, temp_f;

temp_c = CircuitPlayground.temperature();

temp_f = CircuitPlayground.temperatureF();

Serial.println(temp_c);
Serial.println(temp_f);
``` |
| Light sensor value | ```
Serial.print("Light sensor: ");

Serial.println(CircuitPlayground.lightSensor());
```
Values range from 0 to 1023; 300 is common for indoor light. |

| LIBRARY FUNCTION | ARDUINO CODE |
| --- | --- |
| Play a tone on the speaker (fixed duration) | Parameters:<br><br>frequency (*16-bit integer*)—The frequency of the tone in Hz<br><br>duration (*16-bit integer*)—The duration of the tone in milliseconds<br><br>wait (*Boolean*)—The wait duration length after tone in millisecond<br><br>`// Play a 440 hertz tone for`<br>`1 millisecond`<br><br>`CircuitPlayground.playTone(440, 100, 0);` |
| Stop playing a tone previously started with playTone | `CircuitPlayground.speaker.end();` |
| Get accelerometer readings | `float x, y, z;`<br><br>`x = CircuitPlayground.motionX();`<br><br>`y = CircuitPlayground.motionY();`<br><br>`z = CircuitPlayground.motionZ();` |
| Set accelerometer range | `CircuitPlayground.setAccelRange(value);`<br><br>Replace value with one of the following: LIS3DH_RANGE_2_G (smallest but greatest precision); LIS3DH_RANGE_4_G (middle, middle precision); LIS3DH_RANGE_8_G (largest, least precise) |
| Detect the board being tapped | `uint8_t c, clickThreshold, tapValue;`<br><br>`CircuitPlayground.setAccelTap(c, clickThreshold);`<br><br>`tapValue = CircuitPlayground. getAccelTap();`<br><br>Set c to 1 to detect single taps only and to 2 to detect double taps as well. clickThreshold sets sensitivity (0–255). tapValue will be set to 0 (no tap), 1 (single tap), 2 or 3 (double tap). |
| Read the onboard microphone | `int x;`<br><br>`x = CircuitPlayground.soundSensor();`<br><br>Values around 0 are silent; –500 to 500 are loud. |

| LIBRARY FUNCTION | ARDUINO CODE |
| --- | --- |
| Get color sensor values | `uint8_t r, g, b;`<br>`CircuitPlayground.senseColor(r, g, b);`<br>Each color value range is 0 minimum to 255 maximum. |
| Set NeoPixel LEDs | Set the color of NeoPixels.<br>Use a hexadecimal value in the format 0xRRGGBB Hex value 0x00 to 0xFF (0 to 255):<br>`CircuitPlayground.setPixelColor(9, 0xFF0000);`<br>Pixels are numbered counterclockwise from 0 through 9.<br>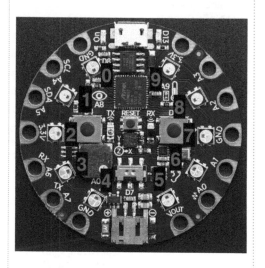<br>The function can also be called with each red/green/blue value:<br>`CircuitPlayground.setPixelColor(9, 255, 0, 0);`<br>To set the NeoPixel brightness for all pixels (0 to 255; default is 30):<br>`CircuitPlayground.setBrightness(50);`<br>To turn off all pixels:<br>`CircuitPlayground.clearPixels();` |

| LIBRARY FUNCTION | ARDUINO CODE |
|---|---|
| Pads A1 through A7 being touched | The touchpads are around the edge of the board. |

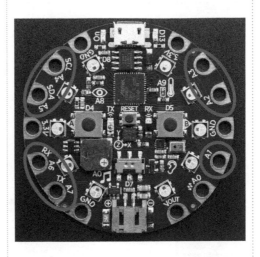

```
if( CircuitPlayground.readCap(A1,
samples) ) {
    Serial.println("Touched pad A1");
}
```

The value samples is optional and defaults to 10. Change A1 to A2, A3, A4, A5, A6, A7 for other pads.

EXAMPLE CODE

Unlike with MakeCode and CircuitPython, Arduino may be best learned by reviewing examples, and using existing code is often encouraged. This is especially true in open source software. Just make sure that if you use someone else's code, you attribute the author, usually in a comment.

The first programs for Circuit Playground Express you should review are the example programs provided by Adafruit. The example files are in the Arduino IDE under File → Examples → Adafruit Circuit Playground (Figure 8-17).

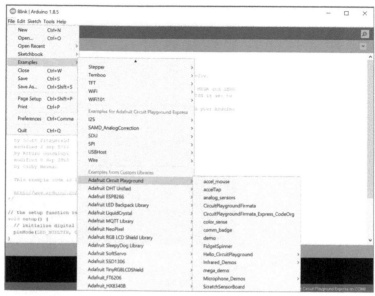

FIGURE 8-17. Finding the Circuit Playground examples in the Arduino IDE

Again, there is no distinction (for the most part) between Circuit Playground Classic and Circuit Playground Express in the example code and when using the library. Most demonstration programs are coded to work with both boards.

But there are a few small differences in the example code. Circuit Playground Express has hardware such as the infrared sensor and LED that the Classic does not. Also, the boards use different microphones.

Study how certain examples like demo and mega_demo go about demonstrating functionality on Circuit Playground Express. If you wanted to implement a project similar to some of the demonstration functionality, consider how you might take the demo code, remove what is not needed, and add what you want.

Example Projects on the Adafruit Learning System

Some of the Arduino Circuit Playground Express examples refer to projects. Those projects and many more are fully described in the Adafruit Learning System.

The Adafruit Learning System has examples for most of their products. They have a special section for Circuit Playground Express at *https://learn.adafruit.com/category/express*. The examples may be for MakeCode, CircuitPython, or Arduino.

Browse the examples and look for the Arduino examples to see what's available.

Your Turn: Arduino Examples

Here are suggestions for your use of Circuit Playground Express with the Arduino IDE:

* Explore the example code in the Arduino IDE for Circuit Playground. Some code is more interesting than others, as some projects assume there is additional building or attachments to a project. Take two demonstration programs, compile them in the Arduino IDE, and load them onto Circuit Playground Express.

* Review the examples in the Adafruit Learning System that use the Arduino IDE. Pick one to review how the author designed the project and the code. Note how the program used library functions to implement device-specific functionality.

LIBRARIES AND COMPATIBILITY

A big allure of programming in Arduino is the hundreds of code libraries that have been written over the past 10 years. These libraries were developed to interface to a wide variety of hardware: sensors, displays, networking, storage, and more.

A good selection of libraries is available in the Library Manager discussed earlier. Others are only available via a personal website or GitHub repository.

If you have a library you have found for one particular project, there is a method for using the code. In the disk folder you are using to save your code, create a subdirectory called `libraries`. Place your library code in that folder. The Arduino IDE should check that folder first before looking in the Library Manager. For example, if your Arduino sketch is in a folder named `myproject`, a library named `awesome` in the local library would be in a folder called `myproject/libraries/awesome`.

NOTE More information on library installation is available at *www.arduino.cc/en/Guide/Libraries*.

Library Compatibility Issues

There is a problem with the vast array of Arduino libraries on the Internet. Some libraries have been coded to take advantage of features on one specific microcontroller. Early Arduino hardware used the Atmel (now Microchip Technology) AVR ATmega series of processors. These were 8-bit microcontrollers; the chips had smaller amounts of memory and ran slowly compared to today's processors such as Circuit Playground Express.

To overcome the limitations of the older AVR processors, some programmers coded AVR-specific hacks to access features or speed up portions of their code. For the processor and project they were using, this worked. When new Arduino compatibles came along in intervening years, the library no longer worked. Some libraries were updated, but many were not.

Circuit Playground Express and other products by Adafruit have wide compatibility with frequently used libraries. This is because Adafruit has invested in writing code to support the modern features of their products. Unfortunately, with third-party boards, they may or may not work with Circuit Playground Express.

Working with Third-Party Libraries

If your project relies on a library on the Internet, here's how to see if it will work as you wish.

1. Read the documentation for the library. Look for which boards the library might have been designed for. Look to see when the library was last updated by the author(s). Older, poorly maintained libraries may not have added compatibility for modern processors such as the SAMD series used by Circuit Playground Express and Arduino MKR1000.

2. If the library is usable, download it and place the code files in a `library/libraryname` folder in your project folder on your storage device.

3. In your code, go ahead and code basic library functionality or create a quick test program to use the library. Be sure to `#include` the library.

4. Try to verify and compile the program that uses the library. Were there errors? If you saw errors, you may or may not be able to change the code in your program or in the library to have the program work with Circuit Playground Express.

If you have used due diligence in trying to make the library work and it doesn't, here's how to seek help:

* Check with the library author(s) to ask questions.

* Go to the Adafruit support forum for Circuit Playground Express at *https://forums.adafruit.com/viewforum.php?f=58*.

* Go to the Arduino support forum at *https://forum.arduino.cc/*. Look for a forum specific to your issue.

Be sure in each help location to state that you are trying to use the library with Circuit Playground Express from Adafruit. Also report what specific errors you are seeing.

WRAP-UP

Using the Arduino environment provides some powerful tools for building projects on Circuit Playground Express. The cost for a new user is fairly high:

* Learning Yet Another Programming Language (YAPL)

* Learning how Arduino designs code and learning the basics of the C programming language

* Understanding what functionality standard Arduino libraries provide

* Exploring the Adafruit Circuit Playground library

* Optional: Exploring third-party Arduino libraries for desired functionality

And here are the benefits of using Arduino:

* The code runs very fast.

* The amount of open source software for Arduino is huge.

* Many microcontroller programmers may already know how to program in Arduino.

* Arduino code may be portable across different microcontrollers.

On the last point, as this is a Getting Started series book, most readers are likely not experienced Arduino programmers. This explains why the book focuses on MakeCode and CircuitPython.

If you intend on programming microcontrollers on a regular basis, certainly learning Arduino is a great thing to do. See Appendix B for additional materials to learn Arduino coding and use.

CHAPTER QUESTIONS

1. The Arduino programming language code is based on what classic programming language? Would studying this language help in crafting good Arduino code? Are resources available on the Internet to learn this language?

2. What are the equivalent Arduino code statements for Make-Code `on start` and `forever` loops?

3. No wrong answer: Which programming language would you like to use for programming Circuit Playground Express? One of the three listed in the book? Or do you like another language available?

Troubleshooting

Running into problems and solving them is a defining part of the Maker experience. This appendix will help you resolve many common issues you may face when working with Circuit Playground Express.

Most issues fall into the following categories:

* Cable issues
* Connectivity issues
* Software issues
* Common library problems
* Error messages
* Usage issues
* Manufacturer support

USB CABLE AND POWER ISSUES

> NOTE Many Circuit Playground Express issues are ultimately traced to a bad USB cable or a power issue.

To get an older USB Micro-B cable, you may scrounge in a box of old cables to find something that works. This approach does not always get you the reliable cable you need. Problems you may encounter include the followinsg:

* Some USB Micro-B cables have only power wires and no signal wires (they were designed for charging devices only).

* The cable connections are broken or intermittent, due to flexing (often at one end or the other). This can happen if the cable was heavily used.

* The wire gauge of the cable is insufficient (an uncommon issue, but it might happen with smaller or more inexpensive cables, often sourced from discount suppliers).

* A connector is cracked, dirty, or broken inside.

With power, be sure the green "On" LED is steadily green at all times. With data, you can go into Microsoft MakeCode and create a small test program and see whether Circuit Playground Express will load the code and execute it. Using MakeCode is simplest at this point since it requires no external software other than a web browser and the website *https://makecode.adafruit.com/*. If you do not have Internet connectivity, you can code simple programs in CircuitPython or Arduino to accomplish a similar result.

You may think, "This cable works for my phone—it should be good." However, the phone may not use the data wires per standard USB specifications, or it may have only power wires. That the cable works for a phone is not a sufficient indication that the cable will work 100 percent in Circuit Playground projects.

Here are some power troubleshooting steps:

1. Check your connections and USB port to make sure that everything connects well.

2. If there is a problem, try swapping the cable for a thicker, more substantial one, or consider purchasing a new one.

3. As a final check, disconnect the USB cable and connect your Circuit Playground Express to external power. You have three choices:

   * Connect a charged LiPo battery to the JST battery connector opposite the USB connector.

   * Use a "Phone Rescue" battery, the type that uses a rechargeable battery and a USB-to-micro-B cable, often to provide extra power to a mobile phone or tablet.

   * Adafruit sells a battery pack taking three AAA cells and provides a battery connector suitable for Circuit Playground Express (Adafruit product # 727). Be sure the on-off switch is in the on position.

Buying a good, substantial cable (Adafruit #2008 or similar) from a local shop or reputable online supplier will remedy many issues.

If at this point you have tried to power the board using multiple methods and the power On LED will not glow green and the board appears dead, see the section "Manufacturer Support," later in this appendix.

If you plugged the board in and there was a flash and now it appears dead, wait about 10 minutes and try again with all external connections removed. There could have been a short circuit if metal touched bridged pads on the bottom of the board.

At this point, if you apply proper power and you do not get a green On LED next to the USB port, Circuit Playground Express may be "dead." It can happen to electronics, especially if they are treated poorly. If your Circuit Playground Express is new, contact the manufacturer. If the board has been working for a while and you know you did something to make it no longer work, you may need to get another Circuit Playground Express. It happens at times—it is part of the experimentation process.

CONNECTIVITY ISSUES

Circuit Playground Express may have problems talking to a larger computer used to program the device. First review the "USB Cable and Power Issues" section prior to diagnosing connectivity issues to ensure the problem is not power or cable related.

Problems with connectivity include the following:

* Intermittent communications on USB 3 ports on computers (USB 3 connectors often have blue plastic inside them)

* Compatibility issues on USB ports on some versions of the Linux operating system

* USB ports not recognizing Circuit Playground Express

Connectivity problems generally do not result in error messages. Look at these possible situations:

General communications: Is your USB cable connected to a USB 3 port?

Reconnect your Circuit Playground Express to a USB 2 port if you have one available. If you are using a USB 2 hub, try to plug into the main port and not the hub. If you use a hub, a powered hub would be better to ensure the current available is enough for your project.

I get the green power LED, but my Circuit Playground Express appears to not communicate in any way; my program is not loaded.

Check the "USB Cable and Power Issues" section.

I cannot find Circuit Playground Express in the list of devices in Windows.

In Windows, Circuit Playground Express shows up under the "Unspecified" category of devices (Figure A-1). In Device Manager, it is under Ports (COM & LPT) as a USB Serial Device, with the Windows communications port listed in parentheses (Figure A-2).

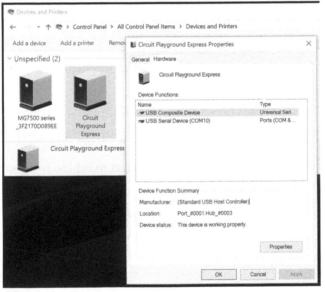

FIGURE A-1. Circuit Playground Express listed in Windows 10 Devices and Printers section of Control Panel

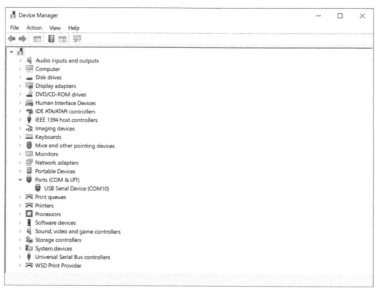

FIGURE A-2. Windows Control Panel Device Manager listing Circuit Playground Express as a USB Serial Device (COM10 here)

The COM port may be numbered differently depending on the port your Circuit Playground Express is plugged into.

I'm using VMware [or another virtual machine program] and I'm having issues.

Some VM programs have problems connecting to "real" computer USB ports. This is not a problem limited to Circuit Playground Express but applies to any USB device. See the VMware or other software forums for USB-specific advice.

Both Microsoft MakeCode and CircuitPython will not work with my Circuit Playground.

Check that you have Circuit Playground *Express*. The difference between the Express and the Circuit Playground Classic is discussed in Chapter 1. Circuit Playground Classic does not run MakeCode or CircuitPython.

My Circuit Playground Express worked when I first got it, but it is acting up now. What could be the problem?

First, check your power connections; if they are not good, correct them. Next, if you have connected external components, your circuit could be electrically problematic or miswired. Remove any connections and try to load a basic Blink sketch in one of the programming languages to test it out. If it works outside your project, check your project connections. If the power On light does not come on, check the "USB Cable and Power Issues" section.

Can I charge a LiPo rechargeable battery to power my Circuit Playground Express?

You can use a LiPo battery for Circuit Playground Express, but note the board cannot recharge the LiPo if it's plugged into a USB port. Unplug the battery from your Circuit Playground Express and charge the battery with a circuit board specifically designed to recharge the battery safely. Adafruit sells several types of LiPo recharging boards; product #s 1304 and 1904 work well at a low cost. The size of the LiPo battery will determine how long the battery will last—the larger the battery, the longer it will power the board. You can save power in a project by using NeoPixels sparingly. Consider using the slide switch to programmatically "turn off" the NeoPixels.

CIRCUITPYTHON ISSUES

I edit my program and copy it to the CIRCUITPY drive but Circuit Playground Express doesn't behave as if it recognizes the code.

The CircuitPython file you copy over to run on Circuit Playground Express should always be called `code.py`. This is so that the Python interpreter knows the name of the code you want to run.

If you use a name like `mycoolprog.py`, the board will not know that file is the code you wish to run. Feel free to use a more descriptive name on your backup storage device copies—for example, `music-on-tilt.py`.

There are four options for filenames for the code the board will run: `code.txt`, `code.py`, `main.txt`, and `main.py`. CircuitPython looks for those files, in that order, and then runs the first one it finds.

Adafruit highly suggests that you use the filename `code.py`.

If your program doesn't seem to be updating as you work, make sure you haven't created another code file that's being read instead of the one you're working on.

My CircuitPython program cannot find the libraries/ modules it needs to work.

CircuitPython looks for library code in the subdirectory `/lib` on the onboard flash drive. Go to Chapter 6 to review the process to copy all of the Adafruit CircuitPython libraries into `/lib`. The libraries use approximately 400KB of space. Even with all Adafruit libraries loaded into `/lib`, there should be plenty of room for other code on the CIRCUITPY drive. See Chapter 6 for more details on installing the Adafruit CircuitPython libraries.

I plug in the Circuit Playground Express and I get a CIRCUITPY drive. But I cannot get the CircuitPython `code.py` to copy over or run when it should.

Double-check if you are using the modified `boot.py` for writing files listed in Chapter 7. If that code is running, first move the slide switch back to file mode. You can then rename `boot.py` to **boot-py.old** on the CIRCUITPY drive and press Reset to get the flash drive back to standard operation.

If you have a file on your Circuit Playground Express named `code.txt`, run it instead of `code.py`. Rename or delete `code.txt` and ensure your code is named `code.py`.

Finally, in a rare event, the flash chip may have an issue. Follow these steps to erase the flash chip and enable normal operation:

1. Type this short program and save it to the CIRCUITPY drive as `code.py`:

   ```
   import storage storage.erase_filesystem()
   ```

2. Click the Reset button on the board to ensure the code is run.

3. Now see if you can copy a CircuitPython program of your choice to the board and have it run.

If you are still having issues, follow the instructions in Chapter 6 on reinstalling the latest version of CircuitPython.

ARDUINO IDE ISSUES

At this point you have gone through the connectivity issues, and everything seems to be working. But you appear to be having errors in the Arduino IDE, either during the compile/verify stage or during upload.

I get several errors when I try to upload a program from the Arduino IDE.

Make sure you've selected Circuit Playground Express in the Tools → Board menu and have selected the proper serial port in Tools → Port. If you switch back to another Arduino-compatible board, change the settings appropriately.

COMMON ARDUINO LIBRARY PROBLEMS

There are many problems you can have with using libraries. The most common library-related error messages take the form "*XXXX* does not name a type" or "*YYYY* not declared in this scope." They mean that the compiler could not find the library. This can be due to any of the following causes:

The library is not installed.

See the steps in Chapter 6 on how to install a library correctly.

Arduino cannot find the library folder.

The IDE will find standard libraries and libraries installed only in the sketch libraries folder.

The specific library folder must be at the top level of the libraries folder. If you put it in a subfolder, the IDE will not find it.

You do not have a "Sketchbook" folder.

It is there, but on a Windows or macOS machine it is named Arduino (on Linux it is named Sketchbook).

A library is incomplete.

You must download and install the entire library. Do not omit or alter the names of any files inside the libraries folder.

A folder name is wrong.

The IDE will not load files with certain characters in the name. Unfortunately, it does not like the dashes in the ZIP filenames generated by GitHub. When you unzip the file, rename the folder so that it does not contain any illegal characters. Simply replacing each dash (-) with an underscore (_) usually works. If the folder has the word master on the end (usually preceded by a dash), remove that also. The best method to see what the library name should be is to look at the sample code to see what the sample expects the library name to look like.

The library name is spelled incorrectly.

The name specified in the #include line of your sketch must match exactly (including capitalization!) the name in the library. If it does not match exactly, the IDE will not be able to find it. The example sketches included with the library will have the correct spelling. Just cut and paste from there to avoid typographical errors.

You have a wrong version of a library or multiple copies of the same library in accessible folders.

If you have multiple versions of a library, the IDE will try to load all of them. This will result in compiler errors. It is not enough to simply rename the library folder; it must be moved outside of the sketchbook libraries folder so the IDE will not try to load it.

One library depends on another library.

Some libraries are dependent on other libraries. For example, most of the Adafruit graphic display libraries are dependent on the Adafruit-GFX library. You must have the GFX library installed to use the dependent libraries. This is true as well for libraries that use I2C that also expect the Wire library.

The IDE needs to be restarted.

The IDE searches for libraries at startup. You must shut down *all* running copies of the IDE and restart before it will recognize a newly installed library.

I found a wonderful Arduino library that does what I need, but when I try to use it on my Circuit Playground Express, I get errors. What can I do?

The library you found was probably coded for other microcontrollers. Those libraries might use large memory spaces or hardware in other microcontrollers, which may not work on Circuit Playground Express. If you understand how the library code works, you may be able to fix some errors yourself. Performing a Google search for the library name may produce pages where others encountered the same circumstance and recoded the library.

Does a library I found on the Internet work with Circuit Playground Express?

Because there are hundreds of libraries out there written by all sorts of people, it may or may not work. Many libraries expect an Arduino Uno and not the larger processor on Circuit Playground Express. But it doesn't hurt to try—see the previous question to proceed.

ERROR MESSAGES

Error messages may fall into the general categories listed here.

Arduino Compilation Issues

Advanced: Can I write code that will compile one way for Circuit Playground and another for Circuit Playground Express in Arduino?

Yes, the Arduino IDE internal preprocessor provides separate definitions for the boards that can be tested. Circuit Playground Classic can be tested using AVR; see the following sample code:

```
#ifdef __AVR__    // Circuit Playground 'classic'
#include "utility/CPlay_CapacitiveSensor.h"
#else
#include "utility/Adafruit_CPlay_FreeTouch.h"
#include "utility/IRLibCPE.h"
#endif
```

Arduino Upload Errors

Be sure you have done the following:

1. Ensure you have a known good USB cable with both power and data lines.

2. Set the Tools → Board menu to Circuit Playground Express.

3. Ensure Tools → Port is set to the communications port that your operating system assigns when you plug in the board. Often the port will say "Circuit Playground Express" next to it, but it might not if things are being balky.

4. If there are still issues, press the Reset button to see if the Arduino IDE will recognize the board.

The Arduino Serial Monitor

While using Arduino, you should use the `Serial.print` and `Serial.println` functions to provide feedback in the Arduino serial monitor as to what the board is doing while coding and debugging. Select Tools → Serial Monitor to see the output (unlike Mu, the output goes to a separate window that you must specifically open after the program is running). The Serial.print statements can later be commented out for a final "ready to use" program.

USAGE ISSUES

You may encounter the following issues while using Circuit Playground Express.

Can other pads besides A1, A2, A3, A4, A5, A6, and A7 do capacitive touch? I would like to have more capacitive touch inputs.

Unfortunately, those are the only pads that work with capacitive touch. A0 does not "do" touch, and neither do the other pads. Consider using two Circuit Playground Express boards to add to the number of touchpads. Adafruit sells capacitive touch expansion boards, but the coding would be complex. It's best to use multiple Circuit Playground Express boards.

Windows 7 (or Windows 8) is not recognizing the board.

See Chapter 2 to learn how to install drivers for Windows 7 and 8. Windows 10, macOS, and Linux do not need drivers.

I would like to try using the Arduino IDE in Linux. What are the pitfalls I need to look out for?

The software may need access to the USB port, but this is controlled by root. You may need to set the USB port for dial-out. Check the Adafruit support forums for Linux issues specific to Circuit Playground Express at *https://forums .adafruit.com/viewforum.php?f=58*.

Will a Circuit Playground Express interface to the hardware I have?

The answer is possibly. Two factors are involved: voltage compatibility and software support.

The input and output pads for Circuit Playground Express are 3.3 volts. The external circuitry should work with a digital output of 3.3 volts. External circuitry should *never* put more than 3.3 volts on an input/output pin because this might damage the board (5 volts on the USB connector is fine, though).

Depending on the function of an external circuit, code will be required to make the circuit function. Sometimes code is easy, or it could be quite complex. It is beyond the scope of this Getting Started book to discuss all the external circuits that can be connected and programmed with the board. You may have to experiment and read up on the subject in other resources.

Are the Circuit Playground Express EAGLE CAD circuit board (PCB) layout files available?

Yes; see *https://github.com/adafruit/Adafruit-Circuit-Playground-Express-PCB*.

MANUFACTURER SUPPORT

Adafruit Industries makes customer service and satisfaction a cornerstone of its business. If you still have problems after troubleshooting, you can visit the Adafruit forums (*https://forums.adafruit.com/*) to describe your situation. The helpful forum moderators will be able to assist with additional troubleshooting.

You'll also find many tutorials on using Circuit Playground Express and other Adafruit products at *https://learn.adafruit.com/*.

After posting to the Adafruit forum, if it is evident your board is defective, Adafruit may replace it (at their discretion). Treat your electronics with care and they should last nearly forever. Just don't spill your drink on it or take it to Burning Man, and then suspect it was a factory fault.

Reference Materials

The main subjects of this book—how to write code in Microsoft MakeCode, how to use CircuitPython, and how to use the Arduino IDE—could each easily be the basis for its own full-length book. In this Getting Started series book, we explored each subject in the space available.

In the following sections, other resources for information are listed for further study. Also consider that new information on the subjects in this book will be published after this book goes to print. Using a search engine of your choice can help if you have exhausted the information in this book and the references that follow.

ON THE INTERNET

The Internet provides a wealth of information. All of the references noted are free to view. Adafruit Industries materials are generally licensed so that you can use the materials any way you want (with attribution).

Circuit Playground Express

* Adafruit Circuit Playground Express Guide: *https://learn .adafruit.com/adafruit-circuit-playground-express*

* Adafruit Customer Support forums: *https://forums.adafruit.com/*

Microsoft MakeCode

* Microsoft MakeCode for Circuit Playground Express: *https:// makecode.adafruit.com*

* The main Microsoft MakeCode site: *https://makecode.com*

* Adafruit Learn Microsoft MakeCode: *https://learn.adafruit .com/makecode*

* Information on the UF2 file format: *https://github.com/ microsoft/uf2*

Python and CircuitPython

* Adafruit Welcome to CircuitPython! *https://learn.adafruit .com/welcome-to-circuitpython*

* Adafruit CircuitPython Essentials: *https://learn.adafruit .com/circuitpython-essentials*

* The Python Software Foundation, Python for Beginners: *www.python.org/about/gettingstarted/*

* The Beginners Guide for Programmers: *https://wiki.python .org/moin/BeginnersGuide/Programmers*

* CircuitPython API Reference: *http://circuitpython .readthedocs.io/en/latest/*

* A list of CircuitPython resources maintained by Adafruit: *https://github.com/adafruit/awesome-circuitpython*

Arduino

* Primary Arduino website: *www.arduino.cc*
* Arduino Language Reference: *www.arduino.cc/reference/en/*
* Arduino Tutorials: *www.arduino.cc/en/Tutorial/*
* C Tutorial: *www.tutorialspoint.com/cprogramming/index.htm*
* Adafruit Circuit Playground Express Guide: *https://learn .adafruit.com/adafruit-circuit-playground-express*
* Adafruit Ladyada's Learn Arduino: *https://learn.adafruit .com/ladyadas-learn-arduino-lesson-number-0*

Chrome OS

* Caret Text Editor: *https://chrome.google.com/webstore/detail/ caret/fljalecfjciodhpcledpamjachpmelml?hl=en*
* Beagle Term Terminal Emulator: *https://chrome.google.com/ webstore/detail/beagle-term/gkdofhllgfohlddimiiildbgoggdpoea ?hl=en*
* YouTube video on using a Chromebook with Circuit Playground Express: *www.youtube.com/watch?v=B-PfKv7DCbc*

PUBLICATIONS

The following resources may help you learn some of the concepts in this book:

* *Getting Started with Arduino, Second Edition*, by Massimo Banzi (co-creator of Arduino)
* *Programming Arduino: Getting Started with Sketches*, by Simon Monk
* Once you've grasped the basics of setting up the Arduino IDE, check out books and other resources on creating

projects based on Arduino. Browsing your favorite technical bookstore will provide a wide range of books and magazines.

* There are many books on learning Python, but it is very difficult to recommend any one title as a companion for a Getting Started book. Most Python books are written for more experienced programmers or extensively use concepts such as object-oriented language constructs unnecessary for beginners.

About the Author

Engineer and Maker Anne Barela is currently a consultant for Adafruit Industries, LLC. She recently retired as a senior Foreign Service officer and security engineer for the U.S. Department of State. Anne is a graduate of Whitman College (mathematics/physics) and the California Institute of Technology (electrical engineering). She has also worked at Hewlett-Packard, the Caltech/NASA Jet Propulsion Laboratory, and Boeing. An avid electronics enthusiast, she started with a workbench and Radio Shack parts in high school. Anne is the author of the book *Make: Getting Started with Adafruit Trinket* (as Mike Barela) as well as tutorials on the Adafruit Learning System at *https://learn.adafruit.com/*.

Index

Milton Keynes UK
Ingram Content Group UK Ltd.
UKHW022341030924
447849UK00007B/40